Aneeya Kumar Samantara • Saty

Metal-Ion Hybrid Capacitors for Energy Storage

A Balancing Strategy Toward Energy-Power Density

Aneeya Kumar Samantara
School of Chemical Sciences
National Institute of Science Education and
Research
Khordha, Odisha, India

Satyajit Ratha
School of Basic Sciences
Indian Institute of Technology Bhubaneswar
Khordha, Odisha, India

ISSN 2191-5520   ISSN 2191-5539   (electronic)
SpringerBriefs in Energy
ISBN 978-3-030-60814-9   ISBN 978-3-030-60812-5   (eBook)
https://doi.org/10.1007/978-3-030-60812-5

This Springer imprint is published by the registered company Springer Nature Switzerland AG
The registered company address is: Gewerbestrasse 11, 6330 Cham, Switzerland

Dr. Aneeya K. Samantara dedicate this work to,
*His daughter **Aayra***
Dr. Satyajit Ratha would like to dedicate this work to,
His Parents, Mrs. Prabhati Ratha and Mr. Sanjaya Kumar Ratha

Preface

Rapid growth in the research and development of clean energy storage techniques, in recent years, have yielded a significant number of electrochemically active compounds/materials possessing enormous potential to facilitate the fabrication of next generation electrochemical energy storage devices such as supercapacitors. Problem, however, remains in achieving a balance among the major factors such as energy density, power density, and cycle life etc. Although significant efforts have been made to create an effective electrical energy storage system that would have an energy density close to a battery and power density close to a capacitor, persistent challenges still lie in combining these two altogether different systems to form a cost-effective and safe storage device. Of the several proposed schematics/prototypes, hybrid supercapacitors have both carbon based EDLC electrode and pure faradic (battery type) electrode, which work in tandem to yield high energy densities with little degradation in specific power. The concept of such hybrid devices and the subsequent implementation of battery-supercapacitor-hybrid (BSH) technologies in heavy electric vehicles at the commercial level suggest the flexibility of supercapacitor technology.

As a promising electric energy storage device, supercapacitors can address several critical issues in various fields of application starting from miniaturized electronic devices and wearable electronics to power hungry heavy automobiles. As discussed earlier, the concept of BSH is highly advantageous and could offer high specific energy, specific power, and excellent life cycle. Depending on the electrode configuration and other controlling parameters, these BSHs can have contrasting performance statistics. BSHs such as metal ion (Li^+, Na^+, Mg^{+2}, etc.) based BSHs, acid-alkaline BSHs, and redox electrolyte based BSHs have shown some of the greatest progress in the field of hybrid energy storage technologies in recent times. BSHs, based on metal ions, are of great interest; particularly Li-ion based BSHs, because of the extreme popularity of Li-ion based batteries and their exemplary commercial success. This Brief is an attempt to discuss the recent progress made in the field of metal-ion based hybrid electrical energy storage devices with emphasis on the effect of different metal ions and other constituent components on the overall electrochemical performance of BSHs through detailed comparison of output performance and longevity (in terms of cyclic stability) including device fabrication cost and safety.

The first part of this book discusses the background of supercapacitor and emergence of metal ion hybrid capacitor. Afterwards, a detailed discussion on the materials developed, electrode configuration and measurement for performance evaluation of metal ion capacitor is presented. This book has been written keeping in mind that the broad readership will include graduate students, academic researchers, and industrial scientists and engineers involved in sustainable energy storage and growth.

Khordha, Odisha, India   Aneeya Kumar Samantara
Khordha, Odisha, India   Satyajit Ratha

About the Book

This book provides a concise overview of a typical battery-supercapacitor-hybrid system, particularly the metal-ion capacitors, which combine a capacitive component with a compatible battery electrode, in order to achieve fast charging capability and high energy density, together with enhanced cyclic stability. Hybrid systems minimise the issues and maximise the performance yield of two complementary systems integrated together, which would otherwise be impossible to achieve in the case of individual component(s). A proof-of-concept lithium-ion capacitor technology, following a hybridisation technique, has recently been witnessed, possessing advantages of both the components and disadvantages of none. Besides lithium, this concept has been extended to some of the other emerging monovalent and multivalent metal-ion systems, e.g., sodium, potassium, calcium, magnesium, zinc, aluminium etc. Since these metal ions have a wide range of chemistries associated with them, here we present a concise overview of the fundamental working model and fabrication techniques of these metal-ion capacitors and also summarise the key components such as electrode materials for both capacitive and battery components along with various types of electrolytic media that have been reported/suggested.

Contents

About the Authors

Aneeya Kumar Samantara, PhD is presently working as a postdoctoral fellow in the School of Chemical Sciences, National Institute of Science Education and Research, Khordha, Odisha, India. Additionally, Dr. Samantara is working as an advisory board member in the editorial board of a number of publishing houses. Recently he joined as a Community Board Member of the journal, "*Materials Horizon*", one of the leading journals in materials science published by the Royal Society of Chemistry, London. He has authored about 23 peer-reviewed international journal articles, ten books (with Springer Nature, Nova Science, Arcler Press, Intech Open, Apple Academic Press, CRC Press), and six book chapters. Several publications are in the press now and are expected to be published soon. He pursued his PhD at the CSIR-Institute of Minerals and Materials Technology, Bhubaneswar, Odisha, India. Before joining the PhD program, he completed his master of philosophy (M.Phil.) degree in chemistry from Utkal University and master of science (M.Sc.) degree in advanced organic chemistry from Ravenshaw University, Cuttack, Odisha, India. Dr. Samantara's research interests include the synthesis of metal oxide/chalcogenides and their graphene composites for energy storage and conversion applications.

Satyajit Ratha, PhD pursued his PhD at the Indian Institute of Technology Bhubaneswar, India. Recently he joined as a Community Board Member of the "*Nanoscale Horizon*", one of the leading journals in materials science published by the Royal Society of Chemistry, London. Prior to joining Indian Institute of Technology Bhubaneswar, he received his Bachelor of Science degree, First Class Honours, from Utkal University and his Master of Science degree from Ravenshaw University. Dr. Ratha's research interests include two-dimensional semiconductors, nanostructure synthesis and their application in energy storage devices. He has authored and coauthored about 23 peer-reviewed international journals, one book chapter, and ten books (with Springer Nature, Arcler Press, CRC Press, Nova Science, Intech Open, Apple Academic Press).

Abbreviations

AIB	Aluminum ion Battery
AIC	Aluminum ion Capacitor
ASCs	Asymmetric supercapacitors
CIB	Calcium ion Battery
CIC	Calcium ion Capacitor
CNT	Carbon nanotubes
CV	Cyclic voltammeter
EDLC	Electrical Double Layer Capacitor
GO	Graphene oxide
HIC	Hybrid ion Capacitor
HOPG	highly oriented pyrolytic graphite
KIB	Potassium ion Battery
KIC	Potassium ion Capacitor
LIB	Lithium ion Battery
LIC	Lithium ion Capacitor
LTO	Lithium titanate
MIB	Magnesium ion Battery
MIC	Magnesium ion Capacitor
NIB	Sodium ion Battery
NIC	Sodium ion Capacitor
rGO/RGO	Reduced graphene oxide
SC	Supercapacitor
ZIB	Zinc ion Battery
ZIC	Zinc ion Capacitor

Chapter 1
Introduction

Abstract In this introductory section, a brief description on the essential features of an ideal electrical energy storage system has been provided, including the necessity and urgency of an efficient, flexible, and long-term storage method that will not only mitigate the overdependency on the fossil fuel based primary energy sources, but also promote and sustain the currently emerging renewable energy resources. A short introduction also mentions batteries and supercapacitors as the potential candidates to provide a long-term and high-performance storage solution at both domestic and industrial scale.

Keywords Energy crisis · Renewable energy · Energy storage · Batteries · Supercapacitors

Storage is an essential feature that has been sustaining each of the life forms on the planet earth, since time immemorial. For example, the enormous amount of energy stored in the form of fossil fuel reserves, the rich and vital resources embedded in both the earth's crust and core, the vast pool of nutrients transduced and stored by the plants, and so forth. Man-made or artificial storage is playing its part too, though on smaller scales. We virtually store almost everything, be it memory, food grains, vital information and data, life-saving drugs, and/or vaccines, to name a few. Yet, the most important component that needs storage for the future sustenance of our planet earth is energy. There is no ambiguity regarding the vastness of the energy reserves in the form of fossil fuels, but we must acknowledge the fact that the same have been exploited and consumed by us in a rather unregulated manner for the past few decades (Lampert, 2019). Soon, the reserves would get exhausted, and we can only speculate the chaotic situation that will follow, due to the lack of a robust energy framework that is critical to meet the huge global energy demand that has been on a rapid growth since the last decade. Now, if we look at the available alternatives for the non-renewable energy resources like fossil fuels, the first question that would arise is regarding the kind of energy carrier we would be dealing with. It should be noted that fossil fuels like coal, petroleum liquid, and natural gases are termed as "primary energy resources," whereas electricity is called an "energy carrier." The most popular and convenient of the energy carriers is obviously electricity, which can be generated from various non-renewable and renewable resources (Braun et al., 1991). The second question is related to the characteristic features of the resources we would be using to generate electrical energy. In most of the cases, fossil fuels are used to generate electricity (e.g., thermal power plants, diesel generators). After extraction, fossil fuels are transported to the designated places and stored for the production of electricity as and when required. Therefore, only secondary level storage is required in this case. Renewable energy resources, however, have no directly usable primary energy forms. Besides, they are intermittent and cannot be transported to other places, nor can they be put to work as and when desired. Considering the fact that these renewable resources can generate enormous amount of energy through solar power, wind energy, and hydro power, the same can only be useful if an effective and strong primary level storage framework could be built for the purpose of transportation/transmission. To preserve the already depleted fossil fuel reserves and to prevent further environmental hazards resulting from their excessive consumption, renewable energy resources complemented by a versatile and durable storage technology, at global level, is the need of the hour (Pleßmann et al. 2014).

Let us briefly review the currently available electrical energy storage options at industrial as well as domestic level. Clearly, battery technology captures a major share when it comes to commercially viable storage methods. However, the early forms of batteries (i.e., primary cells) were not built specifically for energy storage, rather they were meant to deliver a steady direct current output by converting the chemical reactions taking place inside the cell into electricity through an external circuit. The concept of storage evolved with the advent of secondary battery cells, as they can reversibly work with the electricity and the chemical reactions taking place inside them, i.e., they can be recharged repeatedly with the help of an external potential difference, which is then stored inside the cell through the chemical compounds/species that hold onto the energy as long as it is not extracted (or discharged) through a suitable load. The secondary batteries based on lead-acid, nickel–metal hydride, and nickel–cadmium technologies are used for the storage, whereas primary cells become useless once the chemical reactions are over, since the reactions that take place are irreversible. These secondary batteries brought radical changes to the concept of electrical energy storage. Even then, batteries do have their share of limitations, arising out of the internal chemistry that runs them. Average cyclic stability, high sensitivity toward thermal aberrations, low specific power, etc. (Pender et al. 2020) are some of the well-known shortcomings of the current battery technologies. And, there is one more major issue with these batteries, and that is their bulky design, which makes them unsuitable for smart and portable electronic devices.

Electrochemical capacitors (also known as supercapacitors/ultracapacitors) can overcome these limitations. They can be charged/discharged for about 1,00,000 cycles or more, tolerate extreme current values, and are least affected by temperature fluctuations (Lu et al. 2008). The early forms of electrochemical capacitors (or supercapacitors) were intended primarily for power applications, because of their extremely large capacitance values and the tendency to charge/discharge rapidly. Since the overall design of a typical supercapacitor device is highly compact owing primarily to the use of electrode/current-collector assemblies having extremely low thickness, therefore, attempts were made to implement them as power sources for the next generation smart, portable, and miniaturized electronic devices, where compactness (space optimization) is one of the most critical factors for commercialization. These supercapacitor devices, however, lack sufficient specific energy, which renders them ineffective for long-term storage (Kim et al. 2015). The commercial interest, in the currently available supercapacitor technology, is limited to power applications only, e.g., start/stop applications, and as a buffer layer between the external power source and the battery stack in heavy electric vehicles. Nonetheless, a lot of interest has been vested in the supercapacitor technology to bring further optimizations/modifications that would make these devices suitable for a wide range of applications, which the current battery technology is unable to address. The only advantage with the current battery technology is the high value of specific energy. Therefore, a significant amount of work is being carried out by researchers to bring the specific energy of supercapacitor devices at par with the batteries. In the coming sections, we would briefly discuss the nature of various types of supercapacitor devices, their origin, methodology, and advantages/shortcomings to get a clear view of the progress made in the past few years, recent trends, and things need to be done in the coming years.

References

Braun GW, Suchard A, Martin J (1991) Hydrogen and electricity as carriers of solar and wind energy for the 1990s and beyond. Sol. Energy Mater. 24, 62–75

Kim BK, Sy S, Yu A, Zhang J (2015) Electrochemical supercapacitors for energy storage and conversion. In: Handbook of clean energy systems. CRC Press, Boca Raton, pp 1–25

Lampert A (2019) Over-exploitation of natural resources is followed by inevitable declines in economic growth and discount rate. Nat. Commun. 10, 1–10.

Lu W, Henry K, Turchi C, Pellegrino J (2008) Incorporating ionic liquid electrolytes into polymer gels for solid-state ultracapacitors. J Electrochem Soc 155:A361. https://doi.org/10.1149/1.2869202

Pender JP, Jha G, Youn DH et al. (2020) Electrode degradation in lithium-ion batteries. ACS Nano 14:1243–1295

Pleßmann G, Erdmann M, Hlusiak M, Breyer C (2014) Global energy storage demand for a 100% renewable electricity supply. Energy Procedia 46:22–31

Chapter 2
Background

Abstract Although batteries are one of the dominant storage technologies in the field of electrical energy storage, they have certain limitations. These limitations, associated with the battery technology, have led to the discovery of the next generation energy storage devices, i.e., supercapacitors. This section provides a concise overview of the limitations of the batteries and how they can be addressed with help of supercapacitors, by forming a combination of both into a hybrid storage module, where they would complement each other in terms of performance, stability, and cost-effectiveness.

Keywords Double layer capacitor · Pseudocapacitor · Asymmetric capacitor · Hybrid energy storage · Energy · Power

Although both batteries and supercapacitors can store electrical energy, they are of contrasting characteristics, with supercapacitors relying primarily on the surface charge adsorption process, while batteries rely upon the charge-transfer process that occurs through the bulk of the electrodes. Since surface phenomena are much faster than the bulk, therefore, supercapacitors have higher specific power than batteries (Burke 2007), while the bulk diffusion process allows batteries to have higher specific energies than the supercapacitors (Burke 2007). Now, the trade-off between specific energy and specific power is rather tricky, and it would be a really tough task to bring a balance between the two. Thus, there are few fundamental bottlenecks to consider: (1) for any ideal electrical energy storage device, both storage period and cycle life should be as high as possible to prevent undesired issues arising out of premature device failure or inefficiency; (2) the concerned storage device should have the ability to quickly modulate itself according to both the energy source and the dissipative load, i.e., it must have the flexibility to absorb maximum power (or current) from the source without any lag and deliver the same as the output according to the requirement specified by the attached load; and (3) it should possess low self-discharge characteristics so as to minimize leakage.

For long-term storage, a device should have a high specific energy value. Currently, we have batteries as the only viable solution. However, cycle stability is an issue with any battery system. The most robust battery system could stay effective for about ~2000 charge–discharge cycles (without affecting the battery chemistry drastically), which is mediocre when we talk about long-term energy storage. Simultaneously, to be able to absorb/deliver large input/output currents, a device needs to possess high specific power. Most of the secondary batteries have moderate specific power values, due to their internal resistance, internal chemistry, and operating temperature. These parameters strictly put the power specifications of any battery device within specified limits, which means batteries would deteriorate or be dead if subject to charge/discharge currents higher than their specified capacity. In this case, supercapacitors can play a significant role owing to their low internal resistance and extremely high specific power ratings. However, the high self-discharge rates in the case of supercapacitor devices render them ineffective for long-term storage and limit their application as supportive structures in battery–supercapacitor modules. There are many instances where both batteries and supercapacitors have been combined to work tandemly, with supercapacitors acting as a strong buffer layer to prevent any overvoltage and/or thermal stress on the coupled battery modules (Burke 2007).

As it appears from the above discussions, batteries and supercapacitors have their share of advantages and disadvantages, and from these we can clearly conclude that none of them can provide us with a long-term or effective storage solution individually. They have similarities and differences as well, based on their internal chemistry, electrode configurations, type of electrolytic contents, performance metrics, etc. Nevertheless, they both can complement each other to form an excellent hybrid system that could help us achieve both high specific power and high specific energy, a combination that could revolutionize the current state of electrical energy storage. The bottom line is how effectively one can built such a hybrid system without affecting the integrity of the individual components and preserving the output and stability of the combination. Since the concept of supercapacitor is relatively new in comparison to the battery technology, thus, a brief account of the current developments in the field of supercapacitor technologies in terms of their fundamental principles, and technological evolutions till date, has been discussed in this book, along with the tools and techniques that have been implemented over the past few years to combine them with battery systems (especially the metal-ion-based systems) to build hybrid storage units and to achieve an optimum performance in terms of specific power and specific energy.

Electrochemical capacitors are generally divided into two broad categories, i.e., electric double-layer capacitor (EDLC) and pseudocapacitor, depending on the charge storage process that occurs in each case. A brief account of the various types of charge storage principles that work in a supercapacitor device have been discussed in the following sections.

2.1 Electric Double-Layer Capacitor

Electric double layer capacitors (EDLCs) are considered as the electrochemical analogues of the conventional dielectric capacitors. The working of a conventional dielectric capacitor is governed by the principle of charge separation through the application of an external potential difference. Here, the charges of opposite nature are separated and accumulated at the respective electrode surfaces through electrostatic charge adsorption process. In the case of EDLC, however, the charge adsorption is triggered through an electrochemical process at the solid/electrolyte interphase and involves the formation of several complex layers depending upon the nature of the interaction between the solid electrode and the electrolytic material. The typical current–voltage response of an EDLC, which is somewhat distorted rectangular or mostly quasi-rectangular in shape, has a close resemblance (not at all similar) to that of a typical dielectric capacitor. Hence, EDLCs can be more appropriately labelled as capacitive rather than capacitors. The reason behind the deviation of the current–potential responses of EDLCs from the ideal rectangular i–V curves of the conventional dielectric capacitors is the presence of mild redox activities at the surface of the electrodes that are primarily composed of carbonaceous base materials with additional functional species attached to the edges and/or basal planes (Yang et al. 2018).

2.1.1 Electrode Materials for EDLCs

The electrode materials, as mentioned in the previous section, are predominantly derived from carbon. The most prolific among them is the graphene, which is a one-atom-thick layer of bulk graphite, usually extracted through several techniques including mechanical (Chen et al. 2012), electrochemical (Su et al. 2011), and thermal processes (Zhang et al. 2013). Besides graphene, various forms of activated carbon are the most commonly used electrode material in EDLCs (Phiri et al. 2019). There are various forms of graphene such as graphene aerogel (Ye et al. 2013), functionalized graphene (Mishra and Ramaprabhu 2011), graphene nanoplatelets, etc. (Han et al. 2013). Apart from these, there are few graphite-derived materials like highly oriented pyrolytic graphite (HOPG) (Foelske-Schmitz et al. 2010), graphite oxide (Gao et al. 2011), graphite paper, etc. (Ramadoss et al. 2017). Besides these, there are other variants of carbon such as carbon nanotubes (CNTs) (You et al. 2013), fullerene (Bairi et al. 2019), carbon fiber/cloth (Cheng et al. 2011), biochars (carbonaceous materials derived through charring of bio-waste materials such as banana peels, coconut coir, etc.) that have also been reported to have good EDLC characteristics (Usha Rani et al. 2020).

It is to be noted that obtaining graphene, experimentally, is rather difficult, and thus the reduced form of graphite oxide (or graphene oxide, i.e., GO), known as the reduced graphene oxide (rGO or RGO), is used in most of the experimental analyses. One more drawback of graphene is its perfectly flat two-dimensional surface, which makes it highly hydrophobic and less suitable for adsorption of atoms, molecules, or dopants. This is taken care by the use of GO as a base material in most of the experimental techniques. The fact that rGO contains few functional groups of its own in the form of hydroxyl, epoxy, or carboxylic groups explains quite well the quasi-rectangular i–V curves obtained from its electrochemical characterizations in a supercapacitor cell. Having said that, the redox activities due to those functional groups/species are hard to distinguish from the overall electrochemical response of rGO. It is due to the fact that both the double-layer formation and the chemical adsorption of the ions on the electrode surface take place almost simultaneously and instantaneously, and even if there is a lag between the two, it is quite difficult to notice the differences due to the narrow time scale. Nevertheless, the redox behavior is significantly suppressed at higher potential sweep rates as the surface redox reactions are not fast enough to synchronize with the rapidly varying potential steps. However, despite the absence of the redox behavior, it has been observed that the i–V response is not fully rectangular in shape. This can be correlated to the fact that electrochemical capacitors (supercapacitors) are quite different from conventional capacitors in several aspects. Therefore, they have a tendency to deviate from the conventional capacitors in terms of their i–V response, since, in true essence, a capacitor should always store charge through the selective adsorption of charges in an electrostatic manner. However, a large number of research reports have labelled EDLC as capacitors, and therefore, we would be using the term capacitor, for unambiguity.

2.1.2 EDLCs and Conventional Capacitors: Governing Parameters and Storage Performance

The similarities between a conventional capacitor and an electrochemical capacitor can only be drawn in the context of the overall design of the device in each case. Both of the systems use simple electrode architectures and are extremely lightweight, due primarily to the implementation of extremely thin electrodes or electrode/current-collector assemblies. This is possible because both the devices rely on surface adsorption/kinetics, which prevents the use of bulk materials/compounds. However, a conventional electrical capacitor often has a potential difference rating that is many folds higher than that of an electrochemical capacitor. For example, a normal household capacitor can have a breakdown potential of ~220 V, whereas a typical electrochemical capacitor is rated to operate safely within a potential difference of ~2–4 V. This is due to the fact that electrical capacitors use dielectrics (e.g., air, mica, or various other insulators) that have high dielectric breakdown potential values, while the electrolytic materials used in supercapacitors would decompose at much lower potential values. Besides, an electrical capacitor serves a completely different purpose than an electrochemical capacitor does.

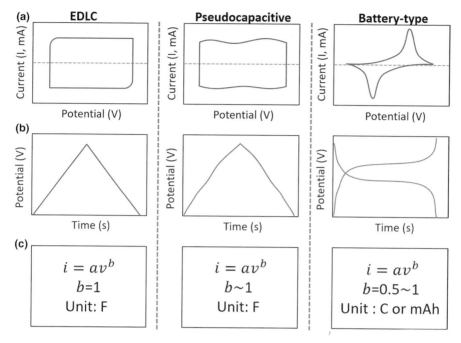

Fig. 2.1 Criterion for distinguishing EDLC, pseudocapacitive, and battery materials. Please refer to the main text for more details. Reused with permission from Jiang and Liu (2019)

Moving to the storage performances of these devices, we can observe a role reversal here, with supercapacitors having extreme capacitance values in contrast to the dielectric capacitors (Fig. 2.1). For example, a typical electrical capacitor would have a capacitance of the order of few microfarads (or at best, few millifarads for capacitors having extremely large dimensions), whereas a typical supercapacitor would have a capacitance of 1 farad or higher. So, basically, a supercapacitor has a capacitance value that is at least 10^3 times greater than the capacitance of an electrical capacitor. Such high capacitance values could be afforded by the supercapacitor devices due to their uniquely designed electrode material possessing an extremely large surface area that allows for the electrochemical adsorption of enormous number of ion species. In addition, the increased capacitance may also be linked to the formation of an electrical double layer at the electrode/electrolyte interphase, resulting in an unprecedented increase in the capacitance value (Zhang and Lee 2014). If we can recall the mathematical expression for the capacitance in the case of a simple parallel plate capacitor, it has two dependencies, i.e., it is directly proportional to the electrode surface area, and inversely proportional to the separation between the two electrodes. The use of carbonaceous materials with high specific surface area (starting from the order of ~1000 m^2 g^{-1}) and a highly compact electrode assembly (using ultrathin membrane separators sandwiched between the electrodes) could impart such huge values of capacitance in the case of supercapacitor.

It is to be noted that, though electrical capacitors can store charge in between its two electrodes due to the electrostatic effect, they were never meant to act as storage devices, since the amount of charge stored is lost within fraction of a second, which obviously cannot be used to do any effective work.

2.1.3 Power and Energy: A Trade-off

Let us invoke the simplest of the mathematical expressions that relates both energy and power through time factor. The concerned mathematical expression somewhat looks like the following:

$$\text{Time} = \frac{\text{Energy}}{\text{Power}} \tag{2.1}$$

Equation (2.1) clearly states that the energy term dominates only when the time parameter is large, which means the power term in the denominator is small. Now, if we interchange power and time, and for the power term to be dominant, the time parameter has to be small. To put things in a simple and straightforward way, power and energy of a system/device are determined by the time parameter and neither of them can be improved without sacrificing the other one. This is one of the many reasons that has motivated most of the research works in designing a system that will have at least a desirable balance between energy and power, especially for energy storage devices.

2.1.3.1 For Capacitors

Now, let us apply this basic idea of energy, power, and time to the dielectric capacitors, EDLCs, and batteries, which will help us develop a clear understanding of their energy–power characteristics. As already discussed, capacitors store energy that is purely electrostatic in nature. The process of electrostatic charge separation is quite fast (much faster than the double-layer formation and reduction–oxidation process), which means they can lose the accumulated charge (on the electrode surface) within a fraction of second (of the order of few milliseconds). According to the energy–power relationship in Eq. (2.1), capacitors, therefore, have highest possible power rating. Thus, capacitors can absorb high input power and discharge the same almost instantaneously (depending on the time constant and the nature of the load attached to the output). However, they are not meant for storing charge, since their capacitance values are considerably low. These capacitors are used for different purposes, e.g., as protective components inside switched power supply modules and for conditioning the power factor in heavy electrical grids (with inductive loads). Unlike the general convention that most of the electrical and electronic appliances require either specified potential difference or current to run properly, the more appropriate term that could be used in this context is power. A stable power input/output means that the product of potential difference (V) and current (I) should possess little or no fluctuation from the rated power of an appliance. This means if there would be a dip in the potential difference, automatically the current would surge. This is more prevalent in the case of digital devices than the analogue ones.

2.1.3.2 For Batteries

Batteries, on the contrary, rely on pure faradic charge-transfer processes. The redox activity mostly penetrates to the bulk of the electrode material and makes the whole process longer in comparison to the surface-based charge transfer. The energy required for the surface kinetics and bulk redox process is then stored as the electrochemical potential energy inside the cell, which can be used as per the requirement. Here, the process of charge transfer that occurs both at the surface and in the bulk of the electrode material is limited by factors like diffusion, intercalation, and chemisorption, etc. The whole process of developing an electromotive force between the two electrodes (i.e., anode and cathode) is not as simple as in the case of a traditional capacitor and takes significant amount of time to complete. This is the reason why capacitors can charge almost instantaneously, whereas batteries could take hours to reach nominal potential values.

2.1.3.3 For EDLCs

Moving to supercapacitors, things are quite different since they can act like both capacitors and batteries, depending on the electrode configuration and internal chemistry. Although we have only discussed about the EDLCs till now, the rest of the discussions on the types of supercapacitors that have been studied so far and their characteristic features in terms of various determining factors would be emphasized in the coming sections. For now, we will stick to the concept of EDLCs, since they were the early forms of supercapacitor technology. Moreover, they have a closer resemblance with the traditional capacitors than the other supercapacitor variants. In the previous section, we have already discussed that EDLCs are capacitor like, and they have nothing to do with electrostatic charge separation process that takes place inside a traditional dielectric-based capacitor. They rely on both the double-layer formation and fast surface redox activities to store charge. Even if there would be no occurrence of any redox activities or charge diffusion, still, the immobilization of charge at the electrode surface due to the electrochemical adsorption and subsequent coulombic interactions will take considerable amount of time when compared to the pure electrostatic phenomenon. Hence, the energy storage is significantly higher in the case of supercapacitors, and of course there is a slight decrement in the specific power. The increase in the storage characteristics when we move from capacitors to supercapacitors can be explained through the following mathematical expression:

$$C = \frac{\varepsilon_0 A}{d} \tag{2.2}$$

where C is the capacitance, ε_0 is the permittivity, A is the surface area of the electrode, and d is the separation between the two electrodes of the capacitor. The expression clearly depicts that there is a direct relationship between capacitance (C) and the surface area of the electrode. At the same time, there is an inverse relationship between the capacitance (C) and the separation between the two electrodes. Hence, either an increase in the value of "A," or a decrease in the value of "d," or both could lead to a noticeable rise in the capacitance value. This is the fundamental working principle based on which supercapacitor devices are able to afford such enormous capacitance values. Traditional capacitors are built with electrodes made from metallic plates/foils with fixed/limited surface areas. Also, the separation between the two electrodes are kept large enough to prevent the dielectric breakdown. EDLCs, on the contrary, use electrodes that have high specific surface areas and also the separation between the electrodes can be minimized to few micrometers. This explains, quite clearly, the enhanced energy storage performance of a supercapacitor device than a conventional capacitor. Furthermore, the tiniest of supercapacitor can have a much higher energy per weight ratio than a capacitor. For example, even the largest of capacitors (having an approximate weight of the order of few kilograms) used in electrical grids can only afford capacitance values as high as few millifarads, whereas a supercapacitor (weighing only about few grams) can have capacitance of the order of few farads. Having said that, EDLC-based supercapacitors are meant for power applications rather than storage as they can lose the stored energy within few minutes (if not seconds, like the case of capacitors). Thus, new concepts such as faradic charge-transfer and bulk diffusion process have been introduced recently. The reason behind the inception of these concepts (which, of course, dilute the capacitive nature of the system and impart extreme non-linear behavior to the whole device) is to slow down the charge accumulation process, which will eventually increase the discharge time window. Further details regarding these new concepts are going to be discussed in the following section.

2.2 Pseudocapacitors

Faradic charge-transfer process (in batteries) and double-layer charge adsorption process (in EDLCs) can be easily distinguished through visual inspection of their respective cyclic voltammetry signatures. However, there is one class of materials that deviates from this and despite having an intrinsic redox-based charge storage characteristic, they produce cyclic voltammetry curves that have striking resemblance to the current-potential response of EDLCs. This unique feature has put these materials under a separate class, i.e., pseudocapacitors. Till now, only two compounds, i.e., MnO_2 and RuO_2, have been recognized as pseudocapacitive, because they can produce nearly rectangular current-potential responses through cyclic voltammetry technique. Nevertheless, there are many literatures reporting several other materials/compounds as pseudocapacitive despite having distinctive redox couple(s) in their cyclic voltammetry responses. Since pseudocapacitive materials rely on redox activities, their specific energy values are higher than that of EDLCs. However, the improvement in the specific energy does not affect the specific power of these materials since their redox reactions take place at the immediate vicinity of the surface or near the surface of the electrode–electrolyte boundary and are quite fast and highly reversible in nature. The underlying mechanism behind pseudocapacitive charge storage has been explained by several researchers. Here, in this book, we have segregated those explanations into distinct categories to address few ambiguities associated with the idea of pseudocapacitive charge storage.

2.2.1 Surface Defects/Irregularities and Overlapping of Redox Couples

When we talk about redox behavior at the surface of an electrode, there has to be some emphasis on the crystallinity and structural orientation of the concerned electrode material. The most common explanation that has been adopted by many reports is based on the assumption that, since a compound cannot have a phase growth that is purely single crystalline in nature, therefore, the possibility of the involvement of the additional phase growths (along with the preferential phase) in the surface activities cannot be discarded entirely. In fact, the polycrystalline nature of these compounds allows for both underpotential and overpotential redox activities besides the standard/primary redox sites (which could otherwise have a distinctive redox signature if the compound would have been single crystalline in nature). A schematic representation has been provided in Fig. 2.2 to invoke a clear picture of the whole mechanism involved.

However, the simultaneous occurrence and overlapping of the redox couples over the whole potential window can only be feasible if the following conditions are satisfied: (1) the surface activities should be fast enough to be indistinguishable from each other, and (2) the faradic charge transfer involving the reduction–oxidation process should be highly reversible, so that a perfect overlapping condition, involving the reduction and oxidation peaks, can be met. This is only possible for elements with high degree of flexibility in their electronic states in order to make available few electrons that would just trigger the charge-transfer process without resorting to any instances of formation and/or breaking of chemical bonds, because if bonds are involved then the process is most likely to lose reversibility and end up producing battery-like behavior rather than being capacitive (Fig. 2.3). This explanation, though hypothetical (since no real-time analyses have been carried out to confirm the same), is somehow able to describe the quasi-rectangular current–potential response from the redox active materials like MnO_2 and RuO_2. Although it might be quite reasonable to explain the pseudocapacitive phenomena through the above analytical statement, it is still not clear enough as to how could the whole process be understood in terms of the well-known Nernst's equation that establishes a well-defined relationship between the electrode potential and the equilibrium coefficient. Nevertheless, the above explanation has long been adopted by most of the researchers in their reports.

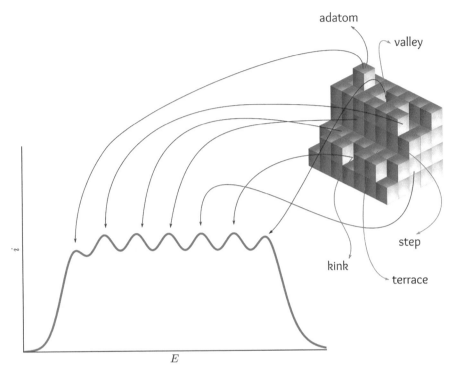

Fig. 2.2 A schematic illustration of common surface defects and their influence on broadening the voltammetric peak via small redox peaks. Reused with permission from Eftekhari and Mohamedi (2017)

Interestingly, and as we have previously stated, there are several reports that have claimed pseudocapacitive behavior for a number of compounds other than MnO_2 and/or RuO_2. These claims have been made on the basis of the fact that besides the intrinsic pseudocapacitance shown by MnO_2 and RuO_2, there are few materials such as the nanostructured layered $Ni(OH)_2$, TiO_2, MoO_3, etc., which have yielded quasi-rectangular i–V responses, because of the phenomena called extrinsic pseudocapacitance. The concept of extrinsic pseudocapacitance is, however, related to the material/electrode designing or modification, especially in the case of nanoscale materials, where the surface-to-volume ratio is extraordinarily high as compared to the bulk. This ratio enables the nanostructured materials to afford much faster surface diffusion processes instead of relatively slower solid-state diffusion that often occurs through the bulk of the electrode material.

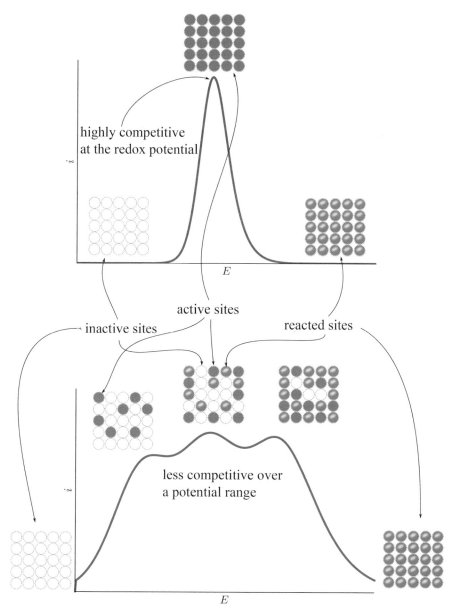

Fig. 2.3 The competition between the neighboring redox sites in battery (top) and pseudocapacitive (bottom) materials. Reused with permission from Eftekhari and Mohamedi (2017)

2.2.2   Explanation in Terms of Band Theory of Solids

If we consider metals, they have a large number of free electrons due to the overlapping of conduction band and valence band, resulting in good electrical conductivity. Due to this property, metals are excellent choices when it comes to electrostatic adsorption (physisorption) of charges in electrical capacitors. However, metals have a rigid surface that cannot be altered, which means their specific surface area remains almost unchanged, therefore no provision for increase in the specific capacitance. Furthermore, since metals have both their conduction band and valence band overlapped, there will be no charge-transfer-based faradic reactions unless bonds are involved. Thus, for redox-based charge storage, metals are considered infeasible due to the fact that they have no capability to hold/isolate a charge. If we move to semiconductors and insulators, they have a finite separation between their conduction band and valence band. Surely, they both have the ability to go under faradic redox reactions; however, the non-capacitive behavior in both the cases can be different since insulators have a comparatively wider and well-defined band gap in contrast to most of the semiconductors (Guan et al. 2016).

Both insulators and semiconductors have active redox sites spread throughout their bulk and the surface as well. Electrode material made from an insulator, when subject to an electrochemical environment, could yield well-defined redox behavior, where the shape and position of the redox couples depend on the nature of the insulator material. If the active sites have energy states that are comparable, then applying an external potential difference would trigger all the active sites to go through a charge-transfer process almost simultaneously. In most cases (for insulators), the active sites are well separated due to strong localized bond structures that do not allow for the conduction of electrons (as evident from the high electrical resistivity in the case of insulating materials), and the oxidation or reduction peak appears only at a certain threshold potential value, for which the active sites simultaneously contribute toward the faradic process (Guan et al. 2016).

Unlike metals, where electrons are highly mobile (not free!) due to the overlapping of the valence band and the conduction band, compounds do not, in general, have the liberty to loose/gain electrons under normal conditions. In electrochemistry, one of the determining factors that govern the flow of electrons through an external circuitry is the mobility of the ions inside the electrochemical cell. Now, the electrochemical cell consists of several non-ohmic components and is highly non-linear in nature, a system completely different from what we observe in the case of a simple metal, where a potential difference would trigger the movement of the highly mobile electrons through a phenomenon called drifting. The amount of the electron flows through the external circuit in the case of a typical electrochemical cell would depend on the valence and the mobility of the ions inside the electrolytic solvent, since they do not allow movement of electrons through them. Thus, ion mobility is one of the major limiting factors in an electrochemical cell. Since mobility of ions inside the electrolyte plays a significant role in determining the charge/discharge characteristics, the choice of a compatible electrolytic mixture is also critical for any of these concerned energy storage devices.

2.2.3 Origin of the Band-Conduction State

Only recently, the development of a whole new concept of the transition from the insulating state to conducting state has changed the way we used to address the phenomenon of pseudocapacitive charge storage in MnO_2 and RuO_2 systems (Costentin et al. 2017). Although this explanation might appear a bit unconventional, since we are accustomed to the much accepted "traditional concept of the pseudocapacitance," it describes the whole process of fast redox behavior through a principle derived from the standard Nernst's equation and suggests that the accumulation of charge on the electrode surface is indeed due to the electrical double-layer phenomenon (Fig. 2.4). This particular theory (along with excerpts from a number of simultaneous experimental results) has also been used to show the transition from non-ohmic state to an ohmic conducting state that establishes an almost linear current–potential response. It is interesting to note that similar approach has already been made to show such kind of transitions in several other compounds such as CoP, SnO_2, ZnO, WO_3, etc. (Costentin et al. 2019), to explain their electrochemical behavior in different electrolytic and pH conditions. But no attempt was made to correlate the methodology with the concept of pseudocapacitance, which has long been a subject of confusion and debate. Fortunately, one can start with understanding the origin of the band conduction state in metal oxides, which include both intrinsic and extrinsic type of pseudocapacitive materials. This, however, can be extended to other compounds like metal chalcogenides, phosphides, carbides, nitrides, etc. with respective model parameters.

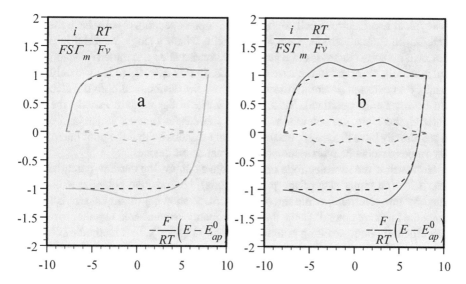

Fig. 2.4 Examples of current–potential CV responses mixing double-layer charging and surface faradaic processes. (**a**) Single faradaic surface reaction with $2(a_Q + a_P - 2a_{PQ}) = F\Delta E°/RT = 2$. (**b**) Two successive faradaic surface reactions with $2(a_Q + a_P - 2a_{PQ}) = F\Delta E°/RT = 1$ for each. Reused with permission from Costentin et al. (2017)

When a certain type of metal oxide (with a well-defined band gap) undergoes an electrochemical reduction–oxidation reaction, the insulating state remains intact until the first oxidation potential is reached. Once the oxidation process starts, the transition from the insulating phase to a conducting state is triggered, and the compound behaves like an ideal metal electrode (similar to the metal electrodes in a dielectric capacitor) facilitating the accumulation of charges at the surface. This is probably the reason why a quasi-rectangular current–potential response is maintained throughout the cyclic voltammetry curve, except for the onset potential that triggers the transition, for both anodic and cathodic scans. However, in most cases, the peaks corresponding to the oxidation (insulating to conducting) and reduction (conducting to insulating) do overlap and are quite hard to visually distinguish during and after the process. This theory clearly demonstrates why compounds like MnO_2 and RuO_2 have EDLC-like i–V responses despite being redox active, whereas the previous explanation did not account for the reversible nature of the pseudocapacitive process despite advocating for a series of rapid and continuous faradic redox couples (Costentin et al. 2017). This is understandable since according to the Nernstian principle, the formation and overlapping of multiple redox couples over a potential range of 0.8–1.0 V will result in an equilibrium constant that would be of extremely high value involving enormous amount of activation energy, which seems, in this case, aptly absurd.

2.3   Asymmetric Supercapacitors

In a typical two-electrode cell configuration, supercapacitors usually have the same material on each electrode. Such arrangement, where a single material is used to fabricate both the electrodes of a supercapacitor, is called a symmetric arrangement and is widely observed in the case of EDLCs. In this case, there is no specific way to apply an external potential difference across the device, and both the electrodes can be reversibly polarized. Since both the electrodes contain exactly the same material, therefore, the potential window in this case is limited, which is of course, as we already have discussed, one of the many reasons behind the poor energy density values associated with symmetric supercapacitor devices.

However, if the two electrode materials have contrasting current–potential characteristics (on either side of the potential axis), i.e., positive potential axis dominated by one electrode, while the other electrode showing dominant characteristic along the negative potential axis, they can probably be combined together to achieve a significantly large working potential window (Fig. 2.5). This method of combining two different electrode materials in a single device to afford a wide potential window and subsequently a high energy density value has led to the concept of asymmetric supercapacitors (ASCs). Although the concept of enhancing the potential window through the asymmetric arrangement of electrodes in a supercapacitor device seems quite promising, the processes involved in finding a compatible set of both the positive and negative electrodes can, in practice, be actually tedious.

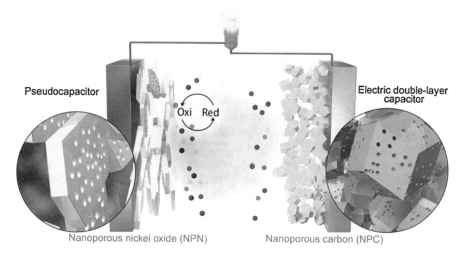

Fig. 2.5 Schematic illustration of ASC based on CPs. NPN derived from cyano-bridged CPs is used as the positive electrode, whereas NPC derived from ZIF-8 is used as the negative electrode. Reused with permission from Salunkhe et al. (2015)

Rigorous material characterizations for both the electrode materials in a three-electrode electrochemical cell are done prior to the device fabrication process to ensure the optimal working potential range for each electrode, for a given electrolytic environment. Once the required optimization steps are completed, the electrodes are combined together in a two-electrode configuration, which is then subject to detailed electrochemical investigation to gain further insights regarding the device performance. Unlike symmetric supercapacitor devices (where there is no distinction between the electrodes), the electrodes in an asymmetric supercapacitor are labelled as positive and negative electrode depending on their respective operating potential windows. Thus, a typical asymmetric supercapacitor will only work properly when connected to an external potential difference by following the correct polarity, changing which could lead to device failure.

From the material perspective, an asymmetric supercapacitor consists of one EDLC-type electrode and another electrode with pseudocapacitive property. Since EDLC-type materials can have versatile potential profiles, i.e., along both the negative and positive sides of the potential axis, they are often used as the negative electrodes. Subsequently, the pseudocapacitive material is implemented as the positive electrode. These pseudocapacitive materials comprise both intrinsic pseudocapacitive materials (e.g., MnO_2 and RuO_2) and extrinsic pseudocapacitive materials (e.g., several binary and ternary metal oxides/sulfides). By introducing fast and reversible redox-based pseudocapacitive technique, the specific energy of these asymmetric supercapacitors can be greatly enhanced, without sacrificing the specific power value.

2.4 Advanced Energy Storage Techniques: Hybrid Metal-Ion Capacitors

The concept of hybrid storage has apparently evolved from the concept of asymmetric supercapacitor devices. The basic idea is to combine a power electrode with an energy electrode (Fig. 2.6). Since the choice of intrinsic pseudocapacitors is too limited (only MnO_2 and RuO_2 are included), therefore, the pool of materials for fast redox-based materials was extended to extrinsic pseudocapacitive materials and pure faradic (battery-type) materials as well. The advantage with these hybrid storage devices is that they can have an improved intermediate cycle life. For example, a secondary battery can run for around 2000 charge–discharge cycles, whereas an EDLC can run for 1,00,000 or more number of charge–discharge cycles. Thus, hybrid devices could remain stable for as high as 5000 to 10,000 or may be a greater number of cycles, without showing any sign of fatigue. However, considering the overwhelming cycle life of EDLCs, we still have to go through a series of evolution and optimization processes to achieve any similar performance metrics in the case of hybrid capacitors. Nevertheless, the concept of hybrid storage has already been taken up by a plentiful of researchers worldwide, and it is only a matter of both time and effort to watch these novel concepts eventually yield desired results.

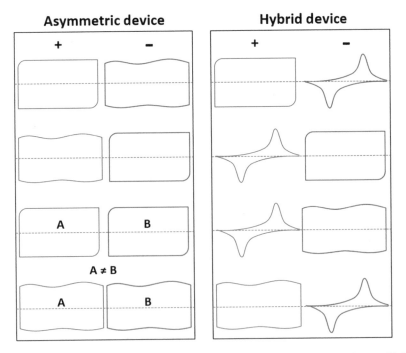

Fig. 2.6 CV features for different device configurations of asymmetric supercapacitors and hybrid supercapacitors (A and B are different materials). Reused with permission from Jiang and Liu (2019)

Having said that, randomly choosing the electrodes and electrolyte could suppress the full potential of each of the electrode and result in device degradation over a considerably short period of time. In order to get close to the excellent cyclic stability of EDLCs, appropriate selection of a cathode material (depending on the type of ion system involved), for hybrid capacitor, is critical, since an incompatible cathode would not only slow down the ion-transport mechanism but also it would degrade much faster with repeated intercalation/deintercalation processes.

In modern battery technologies, most of the dominant secondary storage units are metal-ion based (especially the lithium-ion batteries) with few exceptions like lead-acid batteries. Therefore, for hybridization, metal-ion electrodes are often combined with EDLC materials to construct hybrid storage modules. Here, we will briefly discuss about the metal-ion technologies, their different categories, and subsequent application in hybrid capacitor devices.

References

Bairi P, Maji S, Hill JP et al. (2019) Mesoporous carbon cubes derived from fullerene crystals as a high rate performance electrode material for supercapacitors. J Mater Chem A 7:12654–12660. https://doi.org/10.1039/C9TA00520J

Burke AF (2007) Batteries and ultracapacitors for electric, hybrid, and fuel cell vehicles. Proc IEEE 95:806–820. https://doi.org/10.1109/JPROC.2007.892490

Chen J, Duan M, Chen G (2012) Continuous mechanical exfoliation of graphene sheets via three-roll mill. J Mater Chem 22:19625–19628. https://doi.org/10.1039/C2JM33740A

Cheng Q, Tang J, Ma J et al. (2011) Polyaniline-coated electro-etched carbon Fiber cloth electrodes for supercapacitors. J Phys Chem C 115:23584–23590. https://doi.org/10.1021/jp203852p

Costentin C, Porter TR, Savéant J-M (2017) How do Pseudocapacitors store energy? Theoretical analysis and experimental illustration. ACS Appl Mater Interfaces 9:8649–8658. https://doi.org/10.1021/acsami.6b14100

Costentin C, Porter TR, Savéant J-M (2019) Nature of electronic conduction in "Pseudocapacitive" films: transition from the insulator state to band-conduction. ACS Appl Mater Interfaces 11:28769–28773. https://doi.org/10.1021/acsami.9b05240

Eftekhari A, Mohamedi M (2017) Tailoring Pseudocapacitive materials from a mechanistic perspective. Mater Today Energy 6:211–229. https://doi.org/10.1016/j.mtener.2017.10.009

Foelske-Schmitz A, Ruch PW, Kötz R (2010) Ion intercalation into HOPG in supercapacitor electrolyte – An X-ray photoelectron spectroscopy study. J Electron Spectros Relat Phenomena 182:57–62. https://doi.org/10.1016/j.elspec.2010.07.001

Gao W, Singh N, Song L et al. (2011) Direct laser writing of micro-supercapacitors on hydrated graphite oxide films. Nat Nanotechnol 6:496–500. https://doi.org/10.1038/nnano.2011.110

Guan L, Yu L, Chen GZ (2016) Capacitive and non-capacitive faradaic charge storage. Electrochim Acta 206:464–478. https://doi.org/10.1016/j.electacta.2016.01.213

Han J, Zhang LL, Lee S et al. (2013) Generation of B-doped graphene nanoplatelets using a solution process and their supercapacitor applications. ACS Nano 7:19–26. https://doi.org/10.1021/nn3034309

Jiang Y, Liu J (2019) Definitions of Pseudocapacitive materials: a brief review. Energ Environ Mater 2:30–37. https://doi.org/10.1002/eem2.12028

Mishra AK, Ramaprabhu S (2011) Functionalized graphene-based nanocomposites for supercapacitor application. J Phys Chem C 115:14006–14013. https://doi.org/10.1021/jp201673e

Phiri J, Dou J, Vuorinen T et al. (2019) Highly porous willow wood-derived activated carbon for high-performance supercapacitor electrodes. ACS Omega 4:18108–18117. https://doi.org/10.1021/acsomega.9b01977

Ramadoss A, Yoon K-Y, Kwak M-J et al. (2017) Fully flexible, lightweight, high performance all-solid-state supercapacitor based on 3-dimensional-graphene/graphite-paper. J Power Sources 337:159–165. https://doi.org/10.1016/j.jpowsour.2016.10.091

Salunkhe RR, Zakaria MB, Kamachi Y et al. (2015) Fabrication of asymmetric supercapacitors based on coordination polymer derived nanoporous materials. Electrochim Acta 183:94–99. https://doi.org/10.1016/j.electacta.2015.05.034

Su C-Y, Lu A-Y, Xu Y et al. (2011) High-quality thin graphene films from fast electrochemical exfoliation. ACS Nano 5:2332–2339. https://doi.org/10.1021/nn200025p

Usha Rani M, Nanaji K, Rao TN, Deshpande AS (2020) Corn husk derived activated carbon with enhanced electrochemical performance for high-voltage supercapacitors. J Power Sources 471:228387. https://doi.org/10.1016/j.jpowsour.2020.228387

Yang X, Wang K, Wang X et al. (2018b) Carbon-coated $NaTi_2(PO_4)_3$ composite: a promising anode material for sodium-ion batteries with superior Na-storage performance. Solid State Ionics 314:61–65. https://doi.org/10.1016/j.ssi.2017.11.016

Ye S, Feng J, Wu P (2013) Deposition of three-dimensional graphene aerogel on nickel foam as a binder-free supercapacitor electrode. ACS Appl Mater Interfaces 5:7122–7129. https://doi.org/10.1021/am401458x

You B, Wang L, Yao L, Yang J (2013) Three dimensional N-doped graphene–CNT networks for supercapacitor. Chem Commun 49:5016–5018. https://doi.org/10.1039/C3CC41949E

Zhang F, Zhang T, Yang X et al (2013b) A high-performance supercapacitor-battery hybrid energy storage device based on graphene-enhanced electrode materials with ultrahigh energy density. Energy Environ Sci 6:1623–1632. https://doi.org/10.1039/C3EE40509E

Zhang J, Lee JW (2014) Supercapacitor electrodes derived from carbon dioxide. ACS Sustain Chem Eng 2:735–740. https://doi.org/10.1021/sc400414r

Chapter 3
Metal-Ion Capacitors

Abstract To address the issue of low energy density in traditional EDLCs, one of the carbon-based electrodes is often replaced by a pseudocapacitive electrode material, to form an asymmetric electrode configuration. This asymmetric combination can afford both fast charge adsorption (due to the presence of the EDLC material) and surface diffusion process (due to the presence of a pseudocapacitive material), and improve the overall energy density of the supercapacitor device. This led to the development of hybrid system, where the pseudocapacitive element is replaced by a battery type electrode material, which further enhanced the specific energy of the supercapacitor device. The easiest choice was to introduce a lithium-based battery electrode, since Li-ion system has been dominating the current battery technology for quite a while. This section provides a detailed discussion regarding the progress of hybrid supercapacitor devices based on the lithium-ion technology, and also furnishes a great deal of information regarding the several additional disruptive metal-ion systems that have emerged since the past few decades (either simultaneously or after the implementation of lithium-based batteries), and their role in the development of next generation metal-ion capacitor technology.

Keywords Metal-ion capacitors · Electrodes · Electrolytes · Monovalent metal-ion systems · Multivalent metal-ion systems · Electrode notation · Capacity · Capacitance

The concept of the metal-ion system originated with the evolution of lithium-based energy storage devices in the 1980s. Prior to that, lithium batteries were being manufactured from metallic lithium, which possessed few stiff challenges due to its high sensitivity toward moisture and temperature. Also, rapid dendritic growth was another factor that made the whole research community move to the new and relatively stable and safer lithium-ion system. Lithium-ion systems are faster than bulk metallic lithium and thus have more specific power. Furthermore, the working potential window is slightly higher in the case of lithium-ion batteries than lithium batteries, and it is already known that while the former belongs to the category of secondary energy storage, the latter (bulk lithium) is a primary energy source and also has a higher cost (since they are composed of pure bulk lithium). So, clearly, lithium-ion systems have both economical and electrochemical advantages over the bulk lithium-based secondary storage systems. This is true for other metal-ion systems based on elements like sodium, potassium, magnesium, aluminum, and zinc, which have gathered significant attention from both academia and industries for their futuristic potential as next generation affordable storage solutions for domestic and grid-level applications. The advantage with metal-ion systems is their high degree of tunability and compactness, which is beneficial for portable applications. In the next few sections, we will focus on the various types of metal-ion systems developed till date and their energy storage prospects, especially as an important component in hybrid supercapacitor devices. Besides, several challenges associated with the currently available commercial metal-ion systems have been discussed with possible solutions in terms of novel electrode design and/or device configuration through integrative approach.

Although it is pretty clear that a typical metal-ion capacitor has the privilege of using both the electrochemical capacitor technology (due to the EDLC component as one of the electrodes) and metal-ion-based battery electrode, the working mechanism of the overall system could, in fact, be a lot trickier than it might appear to us. This seemingly ambitious and interesting concept behind achieving both high specific energy and specific power is, however, pretty difficult to realize in practice. There are factors, governing both the electrochemical processes (supercapacitor and battery), which need to be carefully controlled/tuned in order to impart an optimal ion mobility between the two electrodes through a suitable ion-channeling media. Since the ions (in the metal-ion systems), we would be dealing with, are not all the same (considering their source, electronic behavior, size/radius, nature of chemical interaction, compatibility, etc.), therefore a suitable counter electrode (which will be capable of holding the ions successfully and effectively through interstitial trapping method) is highly essential in each of these metal-ion systems. Therefore, all the three major components, i.e., cathode, anode, and electrolyte, need to be investigated thoroughly prior to the development of a practical hybrid energy storage system of both domestic and commercial interest. In this section, we will discuss about the various kinds of metal-ion systems that are gaining interest in recent times and the complementary capacitive component specifically designed for each of those metal-ion systems.

The metal-ion systems that are going to be discussed here have several categorical behaviors, which can put them into different groups; for example, they can be distinguished as monovalent metal-ion systems (e.g., Li, Na, K) and multivalent metal-ion systems (e.g., Mg, Al, Ca, Zn). Also, on the basis of reactivity, metal ions based on Li, Na, K, Mg, and Ca can be grouped together (although Mg does react a bit slower than the rest), while Al and Zn can be put in another group since they are mostly found to be stable under ambient conditions. There has been an additional categorization of metal-ion systems on the basis of availability of the resources/ reserves, geopolitical, socio-economic, and sustainability challenges. We have two groups in this category. Lithium is the lone member of one of the groups because of its scarcity and geopolitical jurisdictions, whereas rest of the metals have been included in the other group (in the non-lithium category) because of their abundance and availability of the resources throughout the globe. While these categories are more or less interconnected with each other, we will stick to the first set of categorizations that has been done in terms of the valency of the concerned metal ion.

Our aim, here, is to discuss the combination of an EDLC-type electrode with a battery electrode (especially the metal-ion-based electrodes that are gaining rapid popularity nowadays), and in the subsequent sections, we will be emphasizing, in detail, the current progress and future prospects of metal-ion-based capacitors (or metal-ion-based hybrid supercapacitors).

3.1 Electrode Notation (Anode, Cathode, Positive/ Negative Electrode)

The demand for portable energy storage has fueled the research on rechargeable batteries, and a significant advancement in this field has been witnessed in recent times with the implementation of all electric vehicles and other modes of transportations. No need to mention that the 2019 Nobel Prize in Chemistry went to the development of Li-ion battery technology. There are a sizable number of articles on battery systems, where the terms anode, cathode, positive electrode, and negative electrode are quite frequently used, without specifying whether the battery (or cell) is in a discharging or charging state. This could lead to an ambiguous and erratic representation of a straightforward charge storage mechanism taking place in the secondary batteries. The commonly used terms such as anode and cathode, are not at all synonymous with positive electrode and negative electrode, respectively, and need to be presented with more clarity in order to get a much better understanding of the working mechanism of a battery device. Since we already know that oxidation always happens at anode and subsequent reduction reaction takes place at the cathode, therefore, the potential values at which a current flow is established through these electrodes are higher and lower than their respective equilibrium states (i.e., they are simply in a polarized state).

Before discharging, the cell stays in a fully charged state, where the anode remains at the highest potential and cathode stays at the lowest potential. While discharging, the potential at the anode drops below its potential at the rest, and due to this potential drop, the concerned anode can be labelled as the negative electrode. Conversely, the cathode potential gradually increases from its lowest potential state and keeps increasing as the potential at the anode decreases. Thus, the cathode becomes the positive electrode. The discharge process continues until the anode potential drops to its minimum, while the cathode potential reaches the highest value, and current flow is ceased. If the cell is now kept in a charging state, the anode potential, which was at the minimum value after being fully discharged, starts increasing, whereas the cathode potential, which was at its highest value after the discharge step, gradually drops. In this case, anode becomes the positive electrode and cathode is labelled as the negative electrode. Thus, during normal use of a rechargeable battery, the potential of the positive electrode, for both charging and discharging states, remains greater than the potential of the negative electrode. However, the role of each electrode is switched during the discharge/charge cycle.

The electrodes should be labelled when a battery is in a spontaneous reduction/oxidation state, i.e., in a discharging state (with no external potential applied across its terminals). All the electrode notations and the use of terms like positive and negative electrodes, in this brief, are for systems that are in a spontaneous discharging state (unless otherwise stated at any point of discussion).

3.2 Capacity Vs. Capacitance

We often come across the terms capacity and capacitance while dealing with batteries and supercapacitors. The mathematical expressions for capacity (in case of batteries) is

$$\text{Capacity} = \left(\frac{n \times F}{3600}\right) A h = \left(\frac{n \times F}{3.6}\right) mA h \tag{3.1}$$

and the specific capacity can be expressed as:

$$\text{Specific capacity} = \left(\frac{n \times F}{3600\beta}\right) A h = \left(\frac{n \times F}{3.6\beta}\right) mA h, \tag{3.2}$$

where "n" is the number of electrons involved (we can take the valency here, for any specific ion, after balancing the electrochemical reaction), "F" is the Faraday's constant, and "β" is the molar mass of the metal/element that acts as the ion source. Equations (3.1) and (3.2) are general mathematical expressions for calculating the capacity and the specific capacity for elements (metal ions, in this case).

Now, for supercapacitors, the expressions for capacitance can be written as follows:

$$\text{Capacitance} = \left(\frac{Q}{V}\right) F = \left(\frac{I \times t}{V}\right) F \tag{3.3}$$

Here, V is the electrochemical working potential window. For comparison purpose, the unit of capacitance can be converted into capacity, using the following expression:

$$\text{Capacity} (F \times V) = \left(\frac{\text{Capacitance}}{3.6} \times V\right) A h \tag{3.4}$$

Equation (3.4) establishes a simple relationship between capacity and capacitance, which can be used to compare the performance of a capacitive electrode with that of a battery electrode, as we will be dealing with both of them.

3.3 Types of Metal-Ion Capacitors

The evolution of supercapacitors from EDLCs to pseudocapacitors and currently studied hybrid systems that comprise an EDLC-type electrode and another electrode, which is purely faradic (battery type) in nature, has moved through a series of rigorous research activities involving cumbersome material synthesis, characterization, device configuration, and optimization steps. The primary aim behind the development of a hybrid supercapacitor is to improve the energy density or specific energy of the device, which will significantly boost the storage property. Moreover, the use of an EDLC-type electrode provides a large surface area, improving the electrode/electrolyte interaction, and helps in the fast ion-adsorption process, which prevents the degradation of the battery-type electrode due to repeated charge–discharge events. A suitable combination of specific power (due to the EDLC electrode) and specific energy (due to the faradic electrode) could drastically enhance the stability and longevity of the hybrid supercapacitor device. There are several advantages with these kinds of system. First, they can be used in places where either specific power, specific energy, or both are required without any reconfiguration of the internal chemistry or electrode architecture. This is not possible with the currently available commercial battery technologies, as they have strict operating regulations beyond which there could be either complete degradation of the battery or catastrophic breakdowns of the components resulting in uncontrolled exothermic reactions. Second, these hybrid systems are extremely temperature tolerant, as they could operate within a much larger temperature window (e.g., some of these devices can charge–discharge within a temperature range of −50 to 80 °C) (Tsai et al. 2013), which is well above the operating temperature range for the market-dominant lithium-ion technology (which operates within a temperature range of −20 to 60 °C) (Ma et al. 2018). Third, the use of an EDLC electrode significantly reduces the overall weight of the hybrid device, since EDLC-type materials are extremely lightweight and have excellent mechanical and electrochemical stability. Besides, the presence of an EDLC material like graphene (which has an extremely high thermal conductivity value) would help reduce the thermal instability in the electrochemical cell by rapidly dissipating the heat generated at extreme operating conditions of the hybrid capacitor.

EDLC materials and their basic working principles have already been discussed in the previous sections. They are primarily made from carbonaceous materials and have intriguing electrochemical properties. Since EDLC-type materials and their physico-chemical properties and capacitive characteristics have already been well documented in numerous literature reports, here, we will be focusing primarily on the various types of metal-ion systems that have been explored to date and will discuss the possibilities on their integration as high-performance faradic electrodes with the EDLC-type materials, for hybrid supercapacitor devices.

There is no surprise that the list of all the metal-ion systems starts with the most popular lithium-ion technology, which has been dominating the secondary energy storage platform since the past few decades. However, the sodium-ion system is actually a much older concept than the current lithium-ion batteries. Clearly, almost all the metal-ion concepts have been overshadowed by the unprecedented success of lithium ion over the years. Only recently, they are being revived and the interest (at both academia and industrial level) is growing at a rather brisk pace. There are several constraints for the lithium-ion technology, which include cost and recyclability, that have brought back the focus on other metal-ion concepts, derived from earth-abundant elements. Nevertheless, lithium ion still remains the foremost choice when it comes to hybrid energy storage modules, not only because of the light-weight nature of the lithium metal but also due to the high compatibility of lithium ions with the intercalation/de-intercalation process, a feat yet to be achieved by any of those other metal-ion systems we have just mentioned. However, these emerging metal-ion-based charge storage techniques have the potential to yield a strong support structure to provide the lithium-ion systems an extended longevity that it requires, and also, they could be able to boost the future energy storage solution in a significant manner. With just the right amount of research and development, some of these technologies can actually evolve into a leading energy storage option in the coming years. Therefore, each of these metal-ion systems are going to be discussed in detail along with their advantages and shortcomings, which will shed light on several key aspects of the next generation hybrid storage devices. The metal-ion capacitors that we are going to discuss have been listed in the following section, with emphasis on their internal chemistry, electrode materials, and device architectures.

3.3.1   Monovalent Metal-Ion Capacitors

As the heading/title suggests, the redox activity of the faradic electrode in these types of hybrid capacitor is governed by a monovalent ion. The metal-ion capacitors comprising a monovalent ion for the bulk diffusion activity include lithium-, sodium-, and potassium-ion capacitors, which respectively employ Li^+, Na^+, and K^+ ions to move back and forth through the electrolyte media.

3.3.1.1   Lithium-Ion Capacitor

The enormous success acquired by the lithium-ion battery (LIB) over the past few decades is due to the high specific energy associated with lithium. Interesting fact is that there are several metal-ion systems that have theoretical capacities similar to or higher than that of lithium; however in practice, they fail to match the performance

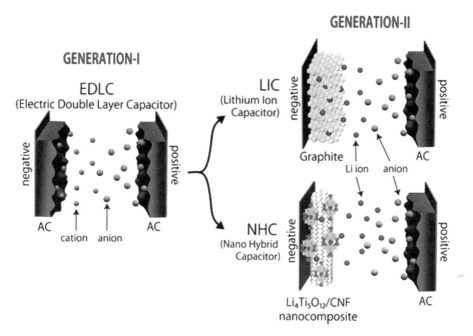

Fig. 3.1 Generation-I is an ordinary electric double-layer capacitor (EDLC) system that has a symmetric cell design utilizing activated carbon (AC) for both their positive and negative electrodes. Generation-II is an Li-ion-based hybrid supercapacitor (Li-HEC) with two typical representative asymmetric configurations. These are the hybrid systems employing a Faradaic Li-intercalating electrode and a non-Faradaic anion adsorption–desorption AC electrode. The Li-HEC is configured with pre-lithiated graphite/LiBF$_4$ or LiPF$_6$(PC)]/AC. The nanohybrid capacitor (NHC)/Li-HEC consists of an AC positive electrode combined with an ultrafast negative electrode made up of nanocrystalline Li$_4$Ti$_5$O$_{12}$, which effectively entangles with the nanocarbon matrix. Reused with permission from Naoi et al. (2012)

of lithium due to sluggish movements of the ions and undesired electrochemical reactions that give rise to larger ionic complexes, drastically affecting both the working potential window and stability of the electrochemical cell. Moreover, lithium-ion technology is the foremost choice when we are concerned about high specific energy because of the large pool of lithium-based cathode materials that are commercially tested and are readily available. The fact that lithium ion is one of the most advanced and refined metal-ion systems currently existing in practice sends a clear and strong message regarding its potential in hybrid charge storage systems.

The concept behind the fabrication of lithium-ion capacitor (LIC) is to have a storage system with an energy density higher than that of a typical EDLC, and a higher power density than that of a lithium-ion battery. LICs employ a pre-lithiated electrode as the negative electrode (anode) and a capacitor-type electrode (e.g., activated carbon) as the positive electrode (cathode) (Fig. 3.1). In a metal-ion battery, the intercalation/de-intercalation process in a layered crystalline framework is governed by the bulk diffusion of the metal ions (here, lithium ion), which significantly limits the charge/discharge rate of the battery device. Supercapacitors, however,

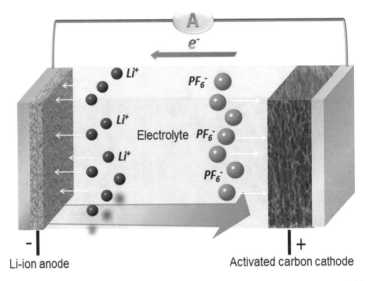

Fig. 3.2 Charging process of a hybrid supercapacitor using AC as the cathode and an Li-insertion material as the anode. Reused with permission from Yi et al. (2014)

store charge mostly through surface adsorption/desorption of ions, where bulk ion diffusion process does not take place. This greatly reduces the capacity of the supercapacitor device. LIC combines the bulk diffusion process (imparts high capacity) and surface charge adsorption (imparts high rate capability) to address low energy density (in the case of supercapacitors) and low power density (in the case of lithium-ion batteries) (Fig. 3.2). The electrolytic media used in the case of LICs are mostly inorganic/organic salts dissolved in suitable organic solvents. Both the electrodes, in combination, facilitate fast and reversible ion-diffusion process over a significantly improved working potential window.

The formation of a solid–electrolyte interface, near the anode surface, significantly reduces the overall capacity of the LIB and requires additional lithium-ion source to compensate the same. Pre-lithiation of the cathode material (which acts as the source of the lithium ion) readily compensates for the irreversible plating of lithium through the formation of solid–electrolyte interface and improves the capacity of the battery device. If the cathode material does not have sufficient amount of lithium ions, then the electrolytic medium provides the additional quantity of the lithium ions, which means the electrolytic content is consumed during the battery operation. Since charge balance is critical for the battery system to work efficiently, calculation of charge–mass balance for anode, cathode, and electrolyte is essential to impart stability to the hybrid storage system.

Fabrication Techniques for Lithium-Ion Capacitors

The early form of lithium-ion-based hybrid storage systems comprised a positive electrode made from a carbon-based material (e.g., activated carbon) and a lithium-based composite (e.g., lithium titanate, LTO) as the negative electrode (anode). This hybrid combination, however, has one major limitation, i.e., the high reversible intercalation/de-intercalation potential of the lithium titanate electrode (vs. Li^+/Li^0), resulting in a low working potential window (Jeżowski et al. 2018). Besides, the specific capacity of the titanate electrode is limited by the process of bulk diffusion and its weight. To address these issues, and to improve the cell potential, lithium titanate was subsequently replaced by a lithiated graphite electrode. The advantages with the graphitic electrode include high specific capacity (almost twice the value observed in the case of LTO), compactness (as both anode and cathode are made from carbon-based compounds, taking less space and volume as compared to LTO), recyclability (the electrodes can be easily recovered once the device is exhausted and recycled to be used again), etc. (Jeżowski et al. 2018). Nevertheless, integrating two devices/systems with contrasting characteristics is not a simple task and requires careful investigation of the underlying principle and working mechanism of the concerned systems at individual level in order to gather critical insights regarding whether the two systems would be compatible or not. Similar case arises with LICs too, and there have been numerous attempts to fabricate LICs (along with other hybrid storage systems) through different approaches (Fig. 3.3). Since the anode (or the negative electrode) in an LIC is not a primary source of lithium ion, additional steps including the pre-charging process required to carry out the lithiation of the graphitic material (in order to implement it as a lithium-ion source) are to be carefully followed before the assembly of the LIC. There are three state-of-the-art methods to accomplish this step, which include the following:

1. *Method—1*: Here, the negative electrode is impregnated with the lithium ions by employing a specifically suited electrolytic media, which acts as primary source of the metal ion. This is achieved by allowing a suitable lithium-based salt, e.g., lithium hexafluorophosphate ($LiPF_6$) or lithium bis(trifluoromethylsulfonyl) imide (LiTFSI), to completely dissolve in a compatible organic solvent, establishing a lithium-ion-rich electrolyte medium. However, the solubility limit prevents the number of lithium ions inside the electrolyte to reach desired levels, so that the lithiation process would not have any significant impact on the ion concentration profile of the electrolytic media. The restricted availability of the lithium ions, thus, leads to depletion of the same during charge/discharge processes, adversely affecting the cyclic stability and the ionic conductivity of the electrolytic media. Furthermore, specific charging profile has to be followed in this case for lithiation process due to diffusion-controlled migration of lithium ions from the electrolyte toward the carbon-based anode. Nevertheless, the gravimetric energy density, in this case, is quite impressive, since only the mass of the active electrode material is taken into consideration here.

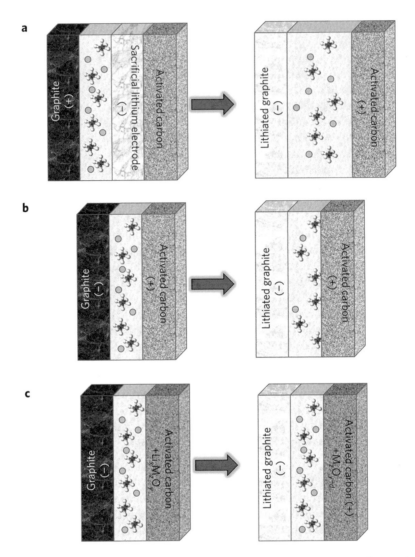

Fig. 3.3 (a) Solution 1 depicts the use of a sacrificial metallic lithium electrode that requires a special preliminary charging step whereby graphite is connected to the negative lithium electrode. In this solution a volume change occurs and must be compensated for prior to further cycling of the LIC; the remaining undissolved metallic lithium is also a potential hazard. (**b**) Solution 2 requires the use of a specially designed charging profile, which provides lithium ions from the electrolyte to the graphite electrode, thus engendering a depletion of positive ions in the electrolyte. (**c**) Solution 3 takes advantage of the overlithiated transition metal oxides, which act as a lithium source on the first charging step to provide lithium cations to the graphite electrode. This needs to occur in an irreversible manner since the oxide must not be electrochemically active on the second and subsequent charge/discharge cycles to ensure optimal performance of the LIC device. Reused with permission from Jeżowski et al. (2018)

2. *Method—2*: This is a patented technology, by Fuji Heavy Industries (Jeżowski et al. 2018), in which an additional electrode made from pure metallic lithium is introduced between the anode and cathode of the LIC system. The lithium electrode serves as a sacrificial lithium source for the anode material through a dedicated discharge step. Although a standard electrolyte comprising a lithium salt is dissolved in an organic solvent ($LiPF_6$, dissolved in a mixture of ethylene carbonate and diethylene carbonate), no part of the electrolyte gets consumed during the lithiation process. There are, however, issues associated with this method. First, using a pure metallic lithium electrode possesses huge safety concerns (as lithium is extremely reactive and may trigger short-circuit and thermal runaway). Second, consumption of the lithium electrode would result in a volume change in the device that can adversely affect the whole electrode/electrolyte/separator arrangement.

   The above issue can be addressed by carrying out the lithiation step separately prior to the assembly of the LIC components. However, this would involve additional steps that will make the whole process both complex and costly.

3. *Method—3*: This is the most recently discovered pre-lithiation technique that makes use of a lithiated metal oxide compound (preferably oxides of transition group metals) as the lithium-ion source for the negative electrode. The lithiated transition metal oxide (e.g., Li_2MoO_3, Li_5FeO_6, and Li_2RuO_3) is first mixed with activated carbon, binder, and conductive additives in appropriate proportions to yield a positive electrode. During lithiation process, the transition metal oxide allows the lithium ions to go through an irreversible intercalation process at the graphitic anode. The successful lithiation process leaves behind the transition metal oxide compound that does not take part in any of the electrochemical activities due to its inert behavior. However, the extraction of lithium ions from these transition metal compounds requires a nominal potential of ~4.7 V (vs. Li^+/Li^0), which can trigger electrolyte decomposition. To solve this, several other lithiated transition metal oxides such as Li_6CoO_4, $Li_{1-x}Ni_{1+x}O_2$, and Li_5ReO_6, etc., have been proposed to bring down the lithium-extraction potential to ~4.5 V (vs. Li^+/Li^0). This method has several advantages over the methods 1 and 2. First, there is no stress on the electrolytic media to provide the lithium ions, resulting in a stable device capacity during charge–discharge cycles. Second, the metal oxide framework is dense, and after the extraction of lithium ions, its volume hardly changes, maintaining a stable electrode/electrolyte/separator architecture. Third, after the lithiation, metal oxide compound remains inert and the added activated carbon acts as the positive electrode facilitating fast charge storage of lithium ion. However, the metal oxide remnant acts as a bulk dead mass and also possesses difficulties in recycling the battery device due to its high density and chemical stability.

4. *Method—4*: The use of an all-inorganic transition metal oxide compound (in method 3) to act as the sacrificial lithium source for the LICs possesses two drawbacks. One is the synthesis process involved, which often takes place at high temperatures. The other one is their large carbon footprint (they are often

synthesized from inorganic base materials like ores by using non-renewable energy resources). For this reason, specially formulated redox-active carbonyl-based electrode materials, including lithium enolate structures, which can be oxidized (delithiated) in the solid state—for instance, tetrahydroxybenzoquinone ($Li_4C_6O_6$), which was used to design the first all-organic and sustainable lithium-ion cells, have been reported recently (Jeżowski et al. 2018). These organic lithium sources can be directly integrated with activated carbon and conductive additives as the positive electrode. Once the lithiation process is over, activated carbon takes over as the anode material for the LIC. This provides a safer, cost-effective, and low CO_2 footprint method for lithiation process and affords a less complex system architecture for LICs.

Similar attempt has been made in a recent report, where the authors have successfully synthesized a novel dienolate sacrificial salt, 3,4-dihydroxybenzonitril edilithium (Li_2DHBN) that is cost effective, is lightweight, and can be prepared via a simple synthesis protocol from its commonly available catechol parent structures. The said organic sacrificial salt yielded a specific capacity of ~365 mA h g^{-1} (Liu et al. 2014).

It is to be noted that LICs are designed to directly compete with EDLCs and asymmetric supercapacitors in terms of energy densities, and not LIBs. In fact, the concept of hybrid ion capacitor (HIC) actually originated from the basic idea of fabricating a supercapacitor device with asymmetric electrodes. The difference between an asymmetric supercapacitor and a hybrid ion capacitor is that the former utilizes a pseudocapacitive material and an EDLC material as the positive electrode and negative electrode, respectively, whereas in a hybrid system, the EDLC material is treated as the positive electrode (cathode) and a compatible battery electrode is used as the negative electrode (anode). Nevertheless, these hybrid systems, e.g., LICs, offer excellent rate capabilities, long cyclic stability, and compactness in comparison to the lithium-ion battery systems. The primary goal behind the concept of LICs is to have a storage system that would have a superior cycle life (than LIBs) and improved energy density (as compared to EDLCs and ASCs).

Below is a summary of advantages of LICs over EDLC and LIBs, which shows that the metal-ion capacitor system indeed provides a safe and improved state-of-the-art energy storage solution for both power and energy applications with greater efficiency and longevity.

Improved Charge Storage Mechanism LICs constitute EDLC-type cathode, which does not have any chemical or thermal instability issues. For anode, LICs use carbon-based materials that undergo a pre-lithiation process prior to being inserted into the hybrid system. Unlike LIBs, LICs are chemically stabilized systems, where the positive electrode does not have to go through any strain effects, since it employs adsorption/desorption process at its surface. The anode (negative electrode) is also made from carbon-based materials, which undergo a pre-lithiation process to accommodate the lithium ions, which lowers the anode potential significantly. This yields excellent charge–discharge characteristics close to EDLCs.

- *Excellent thermal and mechanical stability:* Since there would be no impact on the crystalline framework of the both anode and cathode in LICs, the issue of volume expansion and subsequent effects such as temperature rise and gaseous evolution are significantly suppressed. This provides a huge safety advantage to the LICs over LIBs.
- *Large working potential window:* The working potential window, in the case of LICs, is equivalent to the LIBs (thanks to the battery electrode), which is quite higher than the working potential of typical EDLCs. Also, the capacitance of an LIC is at least twofold higher than that of EDLCs. Therefore, the energy density of an LIC is about fourfold greater than that of an EDLC. This gives a major advantage to the LIC when it comes to accommodating larger energy density values within lesser space (i.e., compactness).
- *Improved short-circuit protection:* In LIBs, when an internal short-circuit occurs, the temperature of the internal cell rises by the short-circuit current. A following reaction between the negative electrode and the electrolytic solution causes an increase in the pressure of the internal cell, followed by a collapse of the crystal at the positive electrode and a release of oxygen in oxidation products of the positive electrode. This causes another thermal runaway, and, in some cases, an ignition or an explosion might occur due to a further rise in pressure of the internal cell and vaporization of the electrolytic solution. This, however, does not occur in the case of LICs. The material composition of their positive electrodes is very different: LIB uses metal oxide, whereas LIC uses carbon-based materials such as activated carbon, which does not contain oxygen. This differentiates their reactions when an internal short-circuit occurs.

The above advantages with LICs are promising and can be exploited in a wide range of applications, both at domestic and industrial scale.

3.3.1.2 Sodium-Ion Capacitor

The intent behind the introduction of sodium-ion capacitor (NIC) immediately after the lithium-ion capacitor is to bring forth the excellent contest between these two metal-ion systems that has been continuing since the past few decades. Interestingly, sodium-ion-based energy storage in the form of sodium-ion battery (NIB) is actually a much older concept than the currently popular lithium-ion battery. However, the rapid popularity gained by LIB along with its improved capacity and fast commercialization pushed the Na-ion system off the edge of the competition. Only recently, the concepts of NIB and NIC have been revisited by the researchers to implement additional metal-ion-based storage methods with a primary aim to promote renewable energy resources.

Sodium-based compounds have several advantages over their lithium counterparts. First, the ionic conduction in the case of sodium β-alumina complexes and sodium superionic conductors (NASICONs) are far more efficient than similar lithium complexes (Tarascon and Simon 2015). Second, sodium-ion batteries have much superior depth-of-discharge profiles than LIBs (Bauer et al. 2018). Besides, NIBs have better thermal stability than LIBs (Bauer et al. 2018). These advantages have provided the Na-ion system a clear edge over other non-lithium metal-ion systems, though strong research activities are being carried out, for other metal-ion systems, at both academic and industrial level to realize the feasibility and effectiveness of these next generation metal-ion-based storage technologies. Currently used electrode materials for sodium-ion systems do not hold great promise, with only a few of them, e.g., activated carbon and sodium-lead alloy, have the possibility of improvement through suitable electrode modification techniques. Low potential difference between the anode and cathode in addition to the sluggish ion diffusion in Na-ion batteries limits the overall charge storage performance. The advantages that Na-ion system has over the Li-ion system include the following:

(a) *Electrode precursor:* Sodium-ion capacitors make use of a broad range of base materials for the preparation of the Na-ion source. These include terrestrial sources in the form of rock salt, sodium carbonate, sodium hydroxide, etc., and aquatic sources such as sea salt or other brackish water reservoirs. These sources have an abundant supply of sodium-based precursors, which can be used to extract sodium in its pure form, or in the form of oxides that can be used in a later stage to synthesize sodium-ion source.

On the contrary, both LIBs and LICs require a sizeable amount of lithium carbonate (Li_2CO_3), which acts as the base material for the preparation of Li-ion source. However, large-scale production of lithium carbonate is a costly and time-consuming process, which is not effective in meeting the commercial target, thereby increasing the cost in the case of both LIBs and LICs.

(b) *Current collector:* The strong affinity of aluminum toward lithium to form alloys at low anode potentials prevents its use as an anode current collector in LIBs (and LICs). The alloying process leads to catastrophic consequences due to volume change in the cell and has been observed even in the case of high anode potential capable lithium titanate anodes. Thus, copper, a costlier and heavier element, is used as the replacement current collector for anode in both LIBs and LICs.

However, for NIBs and NICs, aluminum can readily be used as a current collector for both anode and cathode. This is because aluminum is inert toward sodium. The use of earth-abundant and highly recyclable aluminum would significantly reduce the overall cost of the hybrid storage device.

(c) *Solid–Electrolyte Interphase (SEI):* During lithiation process, in the vicinity of the electrolytic content, a thin layer of lithium gets deposited on the surface of the anode. This is due to the formation of solid–electrolyte interphase, which prevents the anode material to get further oxidized in the presence of the electrolytic content. Similar phenomenon (albeit due to a parasitic oxidation reaction rather than passivation as in the case of SEI) has also been observed in the case of cathode (called cathode electrolyte interlayer, CEI). These can severely affect both coulombic efficiency and capacity of the device. To prevent this, additional steps are followed.

However, this is not the case with sodium-ion capacitors, where the intercalation of the sodium ions into the pores of non-graphitic carbon structures is rare, thus preventing the plating of the sodium species onto the surface of the anode, minimizing SEI formation, averting dendritic growth, and stabilizing both the capacity and the coulombic efficiency of the NICs.

Fabrication Techniques for Sodium-Ion Capacitor

The fabrication of an NIC follows an almost similar process as we have discussed in the case of an LIC. Just like pre-lithiation in the case of LICs, pre-sodiation is an important step toward the fabrication of NICs. The reason behind the implementation of a pre-sodiation step lies with the formation of irreversible solid–electrolyte interphase (SEI) and cathode–electrolyte interlayer (CEI), though their occurrence might be non-significant in the case of NICs. Besides, some of the commonly used sodium salts, e.g., sodium perchlorate (NaOCl$_4$), have very poor solubility in carbonate-based electrolytes. Thus, any sort of irreversible ion loss could drastically affect the conductivity of the electrolytic medium and result in a shorter cycle life of the NIC.

The sodiation process involves a conditioning step in which the anode is cycled against a pure metallic sodium electrode in a half-cell (Fig. 3.4). This process compensates the ion loss due to any SEI/CEI events or irreversible bulk trapping in the case of the anode and keeps its potential close to zero (vs. Na$^+$/Na^0). After the

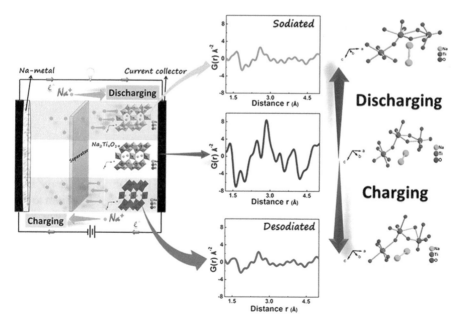

Fig. 3.4 Schematic showing the basic working mechanism of a sodium-ion capacitor. Reused with permission from Bhat et al. (2018)

electrochemical sodiation process, the sodiated anode is then used directly to construct a full-cell hybrid device. Another effective method, which can be used to avoid this extra electrochemical step in a half-cell, is to bring the sodium metal in direct contact with the electrode material, in the presence of the electrolyte. This chemical treatment ensures maximum sodiation of the anode material without needing any additional cell configuration(s). Having said that, such method involving a direct chemical interaction between the anode and the sodium metal is hard to control and determining the final sodium ion count post the chemical reaction would be a daunting task. Nevertheless, this method is easy to accomplish and affords scalability in comparison to the electrochemical method.

Operation Mechanism of NIC

The operational mechanism of a typical sodium-ion capacitor is more or less similar to that of a lithium-ion capacitor. The basic operational principles that an NIC follows are classified into three different categories (Zhang et al. 2020b).

Electrolyte Consuming Mechanism

In this method, the sodium-ion source is the sodium salt (e.g., sodium hexafluorophosphate, $NaPF_6$, and sodium bis(fluorosulfonyl)imide, NaFSI, etc.) dissolved in the organic electrolyte (Zhang et al. 2020b). The anode material, in this case, is composed of carbonaceous material, which is sodiated in the presence of the electrolytic medium through a dedicated charging step. Since a part of the electrolyte (in the form of sodium ions) is consumed here to trigger the sodiation of the anode, the ionic conductivity of the electrolyte and the overall cycle life of the device are significantly reduced.

Na-Ion Exchange Mechanism

In this system, cathode acts as the source of Na ion and anode is made of a capacitor-type material (e.g., activated carbon). During the charge/discharge cycle, the sodium ions are deintercalated from the cathode (unlike the battery, where the ions are deintercalated from the anode) and travel through the electrolytic medium to get adsorbed on the surface of the anode. Here, the electrolyte acts as a mere carrier of the desorbed sodium ions from the cathode and transports them toward the cathode, and the ionic concentration of the electrolyte remains constant throughout the cyclic process. A typical example of this system includes $MXene//Na_2Fe_2(SO_4)_3$, AC//NVOPF@PEDOT systems, etc. (Ding et al. 2015; Dong et al. 2016; Zhao et al. 2018c; Subramanian et al. 2019).

Hybrid Mechanism

This method requires one or both of the electrode materials made from a combination of a battery material and a capacitor material. During the cyclic process, the battery material (which is also a sodium-ion source) at the cathode releases the sodium ions into the electrolytic medium through a desorption process, and the capacitor-type material (already available with the cathode) readily adsorbs the freely available anions from the electrolyte. The desorbed sodium ions from the cathode are then intercalated into the anode. During the discharge process, the anions adsorbed by the cathode are released into the electrolyte and subsequently the sodium ions are deintercalated from the anode to maintain charge balance in the system.

Although sodium-ion capacitors have several advantages, as we have discussed already, it still has a long way to go in order to compete with the rapidly increasing commercial interest in the currently prevalent Li-ion technology. Furthermore, it is hard to integrate high energy density, power density, and appreciable cycle stability in NICs, due to the large capacity/kinetic mismatch between the anode and the cathode. Besides, there are ambiguities associated with the coupling mechanism of anode and cathode, which can hardly be addressed with the help of the limited analytical and experimental results that are currently available. Nevertheless, a robust depth of discharge and better compatibility with the current collector and electrolyte certainly will propel the candidature of sodium-ion-based energy storage technologies in the years to come.

3.3.1.3 Potassium-Ion Capacitor

Limited reserves of lithium and recycling issues have forced the research community to look for cheaper and durable storage technologies, which would be free from lithium-based compounds. In this context, few disruptive concepts such as sodium ion and potassium ion have been discovered and investigated in detail (Fig. 3.5). Although potassium-ion system is naturally assumed to have almost similar or somewhat comparable electrochemical performances with respect to the sodium-ion system, it has been found that the former has, in fact, few key advantages over the latter and also the Li-ion system as well. The following points describe, in brief, several similarities/advantages/limitations of the potassium-ion system as compared to both sodium-ion and lithium-ion systems:

(a) *Availability of precursors:* the availability of potassium in earth's crust is approximately ~20,000 ppm, which is close to the availability of sodium (~23,000 ppm), whereas the amount of lithium availability lurks around 17–20 ppm, which might go up to as high as only ~60 ppm. Another advantage with potassium is that it does not react with aluminum (as is the case with sodium) and can afford aluminum-based current collectors for both anode and cathode in potassium-ion battery (PIB or KIB) and potassium-ion capacitor

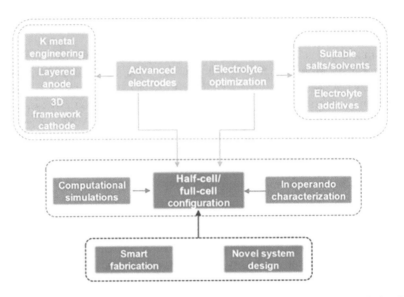

Fig. 3.5 Summary of methodologies in potassium-ion research. Reused with permission from Bhat et al. (2018)

(PIC or KIC). Potassium, in its metallic form, is costlier than sodium, but the potassium salt, K_2CO_3, used to fabricate the anode materials for KIB/KIC comes within the same price range as that of Na_2CO_3 (sodium carbonate, which is the precursor used to fabricate anode materials for NIBs/NICs). This makes K_2CO_3 a much cheaper starting material (or precursor) as compared to Li_2CO_3.

(b) Cell potential: The standard redox potential of K^+/K^0 is −2.93 V (vs. standard hydrogen electrode, SHE), which is lower than the standard redox potential in the case of Na^+/Na^0 (−2.71 V vs. SHE), and is quite close to the redox potential of Li^+/Li^0 (−3.04 V vs. SHE) (Zhang et al. 2019). Therefore, a potassium-ion system would have a cell potential almost similar to that of a lithium-ion system. Interestingly, in the presence of a suitable electrolytic environment, potassium can actually show an even lower potential than lithium, e.g., in propylene carbonate, the redox potential of Li/Li^+ is −2.79 V (vs. SHE), whereas the redox potential for K/K^+ has been recorded at −2.88 V (vs. SHE) (Zhang et al. 2019). Another advantage of potassium over sodium is that potassium ion forms a much more stable intercalation compound with graphite than sodium, which is known to form highly unstable graphite intercalation compounds. Thus, potassium acts as a bridging element between Li and Na, by combining the fast intercalation property of the former and cost effectiveness of the latter.

(c) *Faster ion diffusion:* Among the three alkali metals, i.e., lithium, sodium, and potassium, the latter possesses the largest atomic radius (~0.138 nm), whereas lithium and sodium have atomic radii of ~0.068 nm and ~ 0.097 nm, respectively. However, in a standard carbonate solvent (e.g., propylene carbonate), the Stokes radius for the solvated K^+ is the smallest (~0.36 nm) in comparison to both lithium (~0.48 nm) and sodium (~0.46 nm) (Zhang et al. 2019). This can be explained in terms of the weaker Lewis acidity of K^+ ion, resulting in a relatively smaller Stokes radius in the case of its solvated complex as compared to both Li^+ and Na^+. Thus, potassium ions would have greater mobility than lithium and sodium ions inside a solvent medium, and the resulting higher ionic conductivity is advantageous for both KIB and KIC. Furthermore, it has been observed, from a detailed molecular dynamic study, that the diffusion coefficient of K^+ is about three-fold larger than that of Li^+, imparting faster ion diffusion in the case of the former.

(d) *Choice of current collector*: Similar to sodium, potassium is inert toward metallic aluminum and does not go through an alloying reaction at low potentials as observed in the case of lithium. This eliminates the use of copper as current collector, thereby reducing the overall cost of the hybrid capacitor device.

However, KICs will still deliver lower gravimetric and volumetric energy than LICs due to the larger size and higher mass of K^+ vs. Li^+. It is interesting to note that, since the inception of potassium-ion concept and the realization of first ever laboratory-made KIB (Eftekhari 2004), there has been a sharp rise in the number of reports on KIBs using a wide range of anode and cathode materials. Only recently, the concept of potassium-ion capacitor (KIC) has emerged and holds great promises for cost-effective and high-power capacitive storage along with lithium-ion and sodium-ion systems.

3.3.2 Multivalent Metal-Ion Capacitors

The discovery of lithium-ion capacitor technology was immediately followed by several disruptive hybrid capacitor technologies involving ions of non-lithium metallic elements, and they were not limited to only sodium and/or potassium. Rather, a wide range of elements were immediately tested, which included both monovalent metal-ion systems (Na and K) and multivalent metal-ion systems (Mg, Ca, Zn, and Al). The major blockade in the path of lithium-ion system's large-scale commercial penetration is, of course, the inherent instability associated with the lithium metal, which results in the dendritic growth during anode plating process. Besides, the geopolitical and socio-economic barriers surrounding lithium and its compounds (e.g., Li_2CO_3) are going to turn worse in the coming years due to the rapidly growing energy demand and revolution in storage technologies. The investigation on sodium is obvious, considering its chemical identity with lithium, and the reason potassium came into the competition is because of the strong reaction of sodium with the intercalation cathode materials (e.g., graphite), thereby forming unstable complexes, and also due to the much faster ion-diffusion constant in the case of potassium ion. However, both sodium and potassium possess very low melting point values (~98 °C and ~ 64 °C, for sodium and potassium, respectively), which can be hazardous in the case of thermal instabilities. Thus, implementing both sodium and potassium in their metallic form (even with non-volatile electrolytes) could be challenging. Additionally, the ionic radii for Na and K are

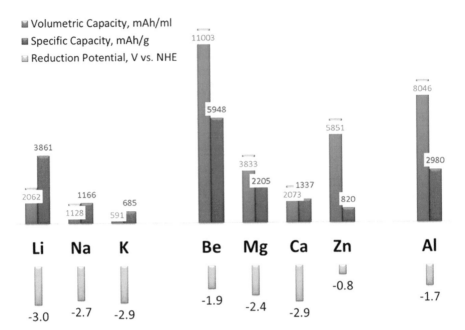

Fig. 3.6 Capacities and reductive potentials for various metal anodes. Reused with permission from Muldoon et al. (2014)

considerably larger in comparison to that of lithium, resulting in slower diffusion kinetics. As the demand for faster and more energy-dense storage technologies is growing at a brisk pace, lithium-ion and other monovalent metal-ion-based storage methods do not have the sufficient capacity to meet the same (Fig. 3.6). Therefore, emerging multivalent metal-ion systems such as Ca^{2+}, Mg^{2+}, Zn^{2+}, and Al^{3+} are currently being investigated. At present, it will be premature to compare these multivalent metal-ion systems with the market-dominant lithium-ion system, as the sole purpose of bringing these multivalent metal-ion concepts is to build a cost-effective and high-performance energy storage environment at global level, for future sustainability.

The elements Li, Na, and K belong to the first group (alkali metals) of the modern periodic table. Since each of them can let only a single electron leave their respective electron cloud to generate the corresponding singly ionized atoms, their energy densities are limited. In contrast, the group 2 metals (i.e., alkaline earth metals), e.g., beryllium, magnesium, calcium, strontium, and barium, can provide two electrons in order to attain a stable electronic configuration. Please note that the element, radium, has been excluded from the list, since much of its chemical properties are still not understood due primarily to the element's high radioactive property. The first element beryllium is a toxic element and is extremely rare in the earth's crust. Both barium and strontium are quite heavy, and if their ionic radii are combined, they do not show promising candidature for electrochemical applications (e.g., metal-ion storage). Magnesium and calcium, on the contrary, are non-toxic and abundantly available. Besides, they are relatively lighter (as compared to barium and strontium) metals, and their ionic radii are surprisingly close to the ionic radii of monovalent metal cations. In fact, the ionic size of magnesium (~0.072 nm) is smaller than that of lithium (~0.076 nm). For calcium, the ionic size (~0.1 nm) is smaller than both sodium (~0.102 nm) and potassium (~0.13 nm). Thus, multivalent metal ions with higher density values and lower ionic sizes have several advantages, especially in comparison to Na and K and therefore have greater potentials in the field of electrochemical energy storage, yet to be exploited on a commercial scale.

Apart from the aforementioned alkali and alkaline earth elements, the concept of zinc-ion-based energy storage (zinc-ion battery, ZIB, and zinc-ion capacitor, ZIC) is gaining gradual momentum. The fact that zinc has a history of over 200 years in the field of electrochemistry (including zinc-based primary and secondary batteries) might have stirred up such a concept to invoke zinc-ion chemistry and check for its potential applicability in next generation energy storage along with other metal-ion systems. Nevertheless, a significant amount of research interest for multivalent metal-ion-based storage technologies has been observed in recent times, which is critical for the future of electrical energy storage.

3.3.2.1 Calcium-Ion Capacitor

Since there are only a handful of reports on calcium-ion-based secondary storage systems, i.e., calcium-ion battery (CIB), calcium-ion capacitor (CIC) is still in a nascent stage. Although the ionic size of Ca^{2+} is very much comparable to that of Na^+, no significant progress has been achieved till date in the case of CICs as compared to NICs. Several key bottlenecks (including several technological limitations) that have hindered the development of CIBs and CICs include the restricted availability of a suitable electrolyte medium, complex plating characteristic, and the extremely limited choice of electrode materials (anode and cathode). It has been found that calcium forms a rather strong surface passivation film (formation of impermeable SEI layer), which does not facilitate the migration of the calcium ion. This is in contrast to lithium, where the SEI provides a favorable ion diffusion path for the lithium ion. This is one of the major reasons behind the poor electrochemical performance of calcium in most of the electrolytic media. Besides, calcium forms stage 1 graphite intercalation compound when allowed to react with a graphite electrode, thus a typical intercalation/deintercalation (as in the case of LiC_6 and KC_8) could not be achieved, further degrading the electrode performance in terms of capacity.

Nevertheless, calcium has several advantages when it comes to cost effectiveness and abundance. It has the highest abundance among all the divalent metal-ion systems (i.e., Ca, Mg, and Zn), and lowest density as well. Wide availability, multiple redox states due to its divalent nature, a standard redox potential of −2.87 V (vs. SHE), which is lower than all the non-lithium metal-ions mentioned here (except for potassium), and fast ion diffusion (due to small charge density), fast reaction kinetics, etc. make it a promising candidate for the calcium-ion battery technology. Besides, calcium metal is less malleable than lithium, which affords the possibility for uniform deposition and avoids the growth of dendrite during charge/discharge cycles. Furthermore, calcium is extremely biocompatible and can help develop one of the safest electrical energy storage technologies available yet. Interestingly, calcium has a higher specific capacity than zinc.

The fabrication techniques and the detailed compatibility statistics (for critical components including current collector, solvent type, salt composition, anode and cathode material, etc.) for CICs require a sound knowledge of the core mechanism behind the electrochemical behavior of Ca^{2+} ion in various electrolytic environments. As we have observed in the case of other metal-ion capacitors, prior information regarding the corresponding metal-ion battery systems is essential in selecting the appropriate base material for constructing the battery-type electrode for CICs. As calcium-ion battery (CIB) is itself a relatively newer concept, the development of CICs would heavily rely on the fact that how fast the CIBs would evolve with time.

3.3.2.2 Magnesium-Ion Capacitor

Among the alkali and alkaline earth metals, magnesium has the highest density, and it has striking similarities with lithium (often referred to as the diagonal relationship between the elements of period 2 and period 3, in the modern periodic table [till boron–silicon], something which has not been fully understood yet) in terms of chemical and electrochemical activities. However, magnesium has a high charge density around its nucleus, which means it would have a strong interaction with the intercalating compound such as graphite. Similar to sodium and calcium, magnesium forms a stage 1 graphite intercalation compound, resulting in severe capacity degradation.

Despite the above issues, magnesium has been found to undergo plating/stripping reaction without forming the passive SEI layer (thus suppressing any dendritic growth). This reduces safety issues while using magnesium metal as anode materials in storage systems, unlike lithium, which is reactive in its pure metallic form. Magnesium-ion batteries (MIBs) are very attractive because magnesium has a lower reduction potential of -2.37 V (vs. SHE), a volumetric capacity (3833 mA h mL^{-1}) nearly twice than that of lithium (2062 mA h mL^{-1}) due to the divalent nature (Muldoon et al. 2014) and relatively low cost originated from the high environmental abundance of magnesium resources. Despite the promising features, the two biggest obstacles encountered by MIBs are the lack of suitable cathode materials in which Mg can diffuse with fast kinetics and the lack of electrolytes compatible with electrodes.

3.3.2.3 Zinc-Ion Capacitor

Zinc has a rich history in the field of battery technology, which extends to at least 200 years. This is due to the fact that zinc, in its pure metallic form, is highly stable in aqueous electrolytic environments, unlike all the alkali and alkaline earth metals that we have discussed in the previous sections. Interestingly zinc has an ionic radius that is smaller than even lithium ion. In addition, zinc metal has several other advantages, for example, it is non-corrosive and has zero toxicity, zero flammability, and the most notable is its compatibility with aqueous solutions. Conversely, all the alkali and alkaline earth metals show very strong reactivity toward aqueous environment and even atmospheric moisture.

The excellent stability in aqueous electrolytic media enables zinc to be directly implemented as a battery electrode in its pure metallic form. During charge–discharge cycles, zinc is plated/stripped and the Zn^{2+} migrates toward the cathode where intercalation/deintercalation occurs. Although zinc has a smaller ionic radius, the sluggish movement of its ions and difficulty in getting adsorbed on the surface of a capacitive electrode (e.g., activated carbon, or graphene) have hindered its implementation in hybrid ion capacitor technology. Also, since it has a very high standard electrode potential (−0.76 V vs. SHE) as compared to that of lithium (−3.04 V vs. SHE), the energy density performance remains poor, and the overall cell potential is drastically limited. The absence of a suitable capacitive electrode material is highly essential for the realization of ZICs. Nevertheless, the high volumetric capacity (5851 mA h cm^{-3}) of zinc in comparison to lithium (2046 mA h cm^{-3}) and magnesium (3833 mA h cm^{-3}) and excellent stability between electrode and electrolyte make zinc a potential contender to realize non-lithium-based multivalent metal-ion capacitors (Muldoon et al. 2014).

3.3.2.4 Aluminum-Ion Capacitor

Metals like aluminum and zinc can be used in their pure metallic form as electrode materials, since they do not react when exposed to moisture or oxygen, though aluminum achieves this through a rapid surface passivation phenomenon, which forms a protective layer at the metal–air interface. Nevertheless, aluminum and zinc both rely on the movement of their respective cations, i.e., Al^{3+} and Zn^{2+}, for charge storage activities. In other words, metal-ion systems are ideal choices for batteries as they can be charged/discharged multiple times and also they are much faster in operation than other rechargeable battery technologies such as sealed lead-acid, Ni–Cd (nickel–cadmium), and Ni–MH (nickel–metal hydride), etc. The electrode structures in the case of reactive metals such as lithium, potassium, sodium, and calcium are different, i.e., their ion sources are found in the form of complex compounds or alloys in order to mitigate the parasitic side reactions that take place between these metals and several electrolytic materials. For both aluminum-ion and zinc-ion systems, the ion sources are primarily the metals in their pure form, as both aluminum and zinc are compatible with a wide range of electrolytic substances.

If lithium has the lowest density among all the metals listed above, aluminum has the second highest density (after zinc). Surprisingly, the ionic size of Al^{3+} is the smallest among all the metal ions and its volumetric energy density is huge (~8046 mA h cm^{-3}), which is almost fourfold greater than that of lithium (Muldoon et al. 2014). Incidentally, the volumetric capacities of both aluminum and zinc are almost comparable. Due to its small ionic form factor and high volumetric capacity, it can be immediately assumed as the best material for the fabrication of high-performance metal-ion capacitors.

However, as we have already observed in the case of CICs, in order to discuss the prospects of aluminum-ion capacitor (AIC), there should be enough know-how support from the corresponding metal-ion battery technology (i.e., aluminum-ion battery, AIB). The concept of aluminum-ion battery started in 2010, and no significant improvement has been made yet. Rather, aluminum ion is still in a confined state due to extreme limitations on the type of electrolyte and cathode material used. Aluminum shares most of its chemical properties with beryllium (due to the diagonal relationship!). Both of them form/prefer covalent bonding, and it is quite hard to control the reaction chemistry of Al^{3+} ion inside both aqueous and non-aqueous environments. On the contrary, Al-ion batteries (that have been reported till date) employ an ionic liquid such as 1-ethyl-3-methylimidazolium chloride ([EMIM]Cl) mixed with plenty of $AlCl_3$. The ion source here is both the $AlCl_3$ salt and the chloride ion available with the ionic liquid. The charge–discharge step involves the formation of chloroaluminate ions (may appear in most commonly observed anionic form $AlCl_4^-$ and $AlCl_7^-$ or in the cationic form such as $AlCl_2^+$), depending on the type of solvent–salt combination.

3.4 Electrodes for Metal-Ion Capacitors

The performance of a metal-ion capacitor heavily relies upon the type of coupling mechanism involving both the electrodes (i.e., anode and cathode), as they form critical components of a metal-ion capacitor, which can have significant impact on the device performance and stability. Since metal-ion capacitors employ a battery (faradic, non-capacitive) type electrode in combination with an EDLC (capacitive, non-faradic) type electrode, we would be discussing, in this section, a number of potential candidates from each category to provide a clear notion regarding the coupling mechanism and the device performance in terms of capacity and working potential window.

3.4.1 Electrode Materials for Lithium-Ion Capacitor

3.4.1.1 Battery-Type Electrode Materials

The specific capacity, cyclic efficiency, and several other parameters governing the performance of a typical LIC depend upon the type of electrode architecture. The following list provides a detailed overview of the various types of compounds/composites that are currently in use or have the potential to be used as effective battery-type electrodes for lithium-ion capacitors. It should be noted that materials like activated carbon can act as both anode and cathode in a hybrid metal-ion capacitor, but the final device performance would depend on the type of coupling (in terms of mass–charge balance and relative electrode potential values) between both the electrodes (i.e., anode and cathode) and the electrolytic content plays a major role too (which will be discussed later).

Metal Oxides

A large number of binary/ternary metal oxides have been reported to have shown appreciable intercalation/deintercalation properties while being used as electrodes in LICs. The advantage with metal oxides is that they are chemically stable and structurally quite dense. Some of the metal oxide compounds may also provide favorable intercalation sites for the metal ions to shuttle back and forth. Metal oxides, especially those containing transition metals, can be categorized into different groups depending upon their intercalation/deintercalation potential in the case of metal-ion storage devices.

- *Titanium dioxide (TiO$_2$):* TiO$_2$ has a reported theoretical capacity of 336 mA h g^{-1}, when it is fully intercalated with the lithium ions (i.e., LiTiO$_2$) (Zhang et al. 2018a). This intercalation process occurs at a potential value of ~1.8 V, which is relatively lower as compared to the decomposition potential of most of the electrolytic materials. However, low electronic conductivity, which is an issue with most of the metal oxides, limits the specific power output of the LICs comprising TiO$_2$ battery-type electrode. To address this issue, TiO$_2$ is often mixed with conducting carbon powders and binding agent (like PVDF) to increase the conductivity. Furthermore, the lithium-ion insertion capacity has also been found to improve in the case of morphologically modified TiO$_2$, where porosity and surface properties were both manipulated in order to reduce the ion diffusion path and improve the strain tolerance. For instance, calcination temperature and structural phase have a distinctive impact on the lithium-ion intercalation property in the case of TiO$_2$. It has been observed that, in the case of bulk, the anatase phase of TiO$_2$ is better suited for Li-ion intercalation as compared to the rutile phase. Also, calcination at high temperature leads to a decrease in the ion intercalation capacity of TiO$_2$. The reported LIC comprising microspheres of anatase TiO$_2$, calcined at a temperature of 400 °C, yielded an impressive power density of 9.45 kW/kg, while possessing a moderate energy density of 31.5 Wh/kg. The capacity retention was found to be 98% after 1000 charge–discharge cycles (Zhang et al. 2018a).

A large lithium-ion diffusion coefficient and ultra-low charge-transfer resistance were observed in the case of urchin-like anatase TiO$_2$, due to its high surface area, which allowed maximum interaction with the electrolytic medium, improving the intercalation/deintercalation process and electronic conductivity. The urchin-like TiO$_2$ produced a power density of 194.4 kW/kg and an energy density of 50.6 Wh/kg, which are quite impressive (Zhang et al. 2018a). Partially replacing Ti^{4+} in TiO$_2$ with other transition metals like molybdenum (Mo) and niobium (Nb) can also improve both the energy density and power density of the doped metal oxide compound. The doped compounds, $Mo_{0.1}Ti_{0.9}O_2$ and $Nb_{0.25}Ti_{0.75}O_2$, yielded energy densities of 41 Wh/kg (at 1.7 kW/kg) and 36 Wh/kg (at 3.2 kW/kg), respectively (Zhang et al. 2018a).

It has been observed that in comparison to the bulk form of TiO_2, nanostructured TiO_2 phases could achieve significantly better lithium-ion insertion/deinsertion kinetics. In fact, nanostructured rutile TiO_2 has shown much improved ion adsorption in contrast to the bulk rutile TiO_2, which has been found unsuitable for Li-ion systems. Figure 3.7 shows three polymorphs of TiO_2, namely, the rutile, anatase, and bronze titania (Bronze-TiO_2) phases, and the corresponding available ion insertion sites for both bulk and the nanostructures. Self-supported arrays of a rutile TiO_2-deposited hierarchical lithium titanate (RLTO) on a titanium foil recorded a discharge capacity of ~142.9 mA h g^{-1} at a discharge rate of 30 °C (Fig. 3.8) (Zhang et al. 2018a). A fabricated LIC comprising RLTO as the battery electrode and a nitrogen-doped carbon nanotube as the capacitive electrode produced an energy density of 74.85 Wh/kg and a power density of 300 W/kg. Bronze-TiO_2 shows even better ion-adsorption capacity. Bronze-TiO_2 nanoparticles grown on multiwalled carbon nanotube (MWCNT) delivered a capacity of 275 mA h g^{-1} at 1 °C and 235 mA h g^{-1} at 300 °C and are a strong candidate to be employed as a high-performance battery electrode for LICs (Zhang et al. 2018a).

- *Spinel-type $Li_4Ti_5O_{12}$:* Being one of the new entrants in the category of lithium titanium oxide, $Li_4Ti_5O_{12}$ has drawn significant attention due to its high coulombic efficiency and extreme tolerance toward strain induced by the intercalation/deintercalation of lithium ions. The reported theoretical capacity of this battery-

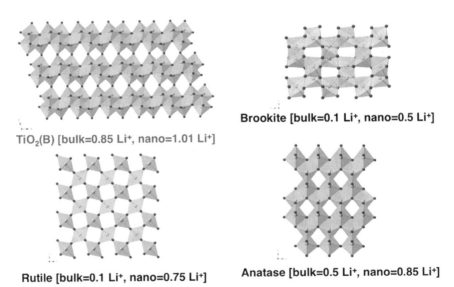

Brookite [bulk=0.1 Li⁺, nano=0.5 Li⁺]

TiO_2(B) [bulk=0.85 Li⁺, nano=1.01 Li⁺]

Rutile [bulk=0.1 Li⁺, nano=0.75 Li⁺]   Anatase [bulk=0.5 Li⁺, nano=0.85 Li⁺]

Fig. 3.7 Unit cells of anatase (**a**), rutile (**b**), and TiO_2(B) (**c**) with idealized lithium-ion insertion sites. Reused with permission from Dylla et al. (2013)

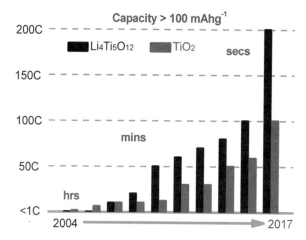

Fig. 3.8 Improvement in the rate capability of TiO$_2$ and Li$_4$Ti$_5$O$_{12}$ Li anodes. The C rates plotted are the maximum values at which >100 mA h g^{-1} of reversible capacity is achieved. Reused with permission from Ding et al. (2018)

type electrode material has been recorded at ~175 mA h g^{-1} (Zhang et al. 2018a). Furthermore, this spinel structure is easy to synthesize and is also cost effective. It has a stable discharge plateau at ~1.55 V, which prevents both the electrolyte decomposition and the formation of solid–electrolyte interphase. However, much like the TiO$_2$, this material also suffers from low ion-diffusion coefficient and poor electronic conductivity. While the ion-diffusion coefficient can be improved with the help of nanostructurization of the material, mixing Li$_4$Ti$_5$O$_{12}$ with highly conducting carbon nanotubes or carbon black can take care of the low electronic conductivity issue. Table 3.1 provides a detailed overview of the performance of composites of Li$_4$Ti$_5$O$_{12}$ with various conducting carbonaceous materials resulting in high conductivity, which is reflected from their impressive power density values.

- *Spinel LiCrTiO$_4$:* This spinel oxide is actually a modified version of Li$_4$Ti$_5$O$_{12}$, where few of the Ti atoms are replaced by the Cr in the crystal lattice. The overlapping of 3d orbitals of Cr with that of Ti results in a high electronic conductivity for LiCrTiO$_4$. Although the theoretical capacity for this spinel oxide is rather low (at ~157 mA h g^{-1}), the discharge plateau is observed at ~1.5, which means both electrolyte decomposition and SEI formation are effectively suppressed.
- *Protonated hexatitanate (H$_2$Ti$_6$O$_{13}$):* A simple ion exchange method is followed to replace Na (or Li) in Na$_2$Ti$_6$O$_{13}$ (or Li$_2$Ti$_6$O$_{13}$) by protons in order to yield H$_2$Ti$_6$O$_{13}$ (Pérez-Flores et al. 2012). The large interlayer separation of H$_2$Ti$_6$O$_{13}$ (HTO) in comparison to other lithium-ion intercalating compounds should, in practice, make it a good battery-type material for LICs. Also, HTO has a high resistance toward strain-induced deformations due to the intercalation/deintercalation of lithium ions. However, it shows a poor cyclic performance. This has been addressed by tuning the morphology and introducing dopant atoms in the HTO lattice.

Table 3.1 Performance of composite material containing $Li_4Ti_5O_{12}$ and conducting material

Battery electrode	Capacitive electrode	Voltage window	Energy density	Power density	Reference
Carbon-coated $Li_4Ti_5O_{12}$	AC	1.5–2.5 V	57 W h/ L 18 W h /L	10–100 W/L 2600 W/L	Jung et al. (2013)
Graphene-wrapped LTO	AC	1.0–2.5 V	15 W h/kg	2500 W/kg	Kim et al. (2014)
$Li_4Ti_5O_{12}$–graphene	AC	1–2.5 V	30 W h/kg	1000 W/kg	Xu et al. (2015)
$Li_4Ti_5O_{12}$–C	AC	1.5–3.0 V	20 W h/kg	37 W/kg	Ni et al. (2012)
LTO–AC hybrid nanotubes	AC	1–2.5 V	90 W h/kg	50 W/kg	Choi et al. (2012)
			32 W h/kg	6000 W/kg	
LTO/CHNS	AC	0–2.5 V	91 W h/kg	50 W/kg	Choi et al. (2011)
			22 W h/kg	4000 W/kg	
Carbonized $Li_4Ti_5O_{12}$	AC	0–2.5 V	330 mW h/L	2.8 kW/L	Vijayakumar et al. (2015)

- *Iron oxides (Fe_2O_3/Fe_3O_4):* Metal oxides often possess large theoretical capacities and low potential plateaus. Theoretical capacities of iron oxides can reach as high as 1000 mA h g^{-1}. However, in practice, their electrochemical performances suffer from their intrinsically poor electronic conductivities. The theoretical capacity of Fe_3O_4 is ~924 mA h g^{-1}, and the conductivity issues can be taken care by using conducting carbon black as an additive or by using a coating mechanism. The biggest advantage with iron oxide-based materials is that they are plentily available and way cheaper than any other metal oxides. LIC based on Fe_3O_4–graphene composite-based battery electrode and capacitive activated carbon yielded an energy density of 204 Wh/kg at a low power density of 55 W/kg and 65 Wh/kg at a power density of 4 kW/kg (Zhang et al. 2013).

  Fe_2O_3 has a theoretical capacity of 1000 mA h g^{-1}, but its performance, again, is marred by low electronic conductivity and strong aggregation property. The process of directly growing Fe_2O_3 nanoparticles on highly conducting substrates like graphene or carbon nanotube can significantly enhance the electronic conductivity. The selective growth of nanoparticles also prevents the aggregation of the Fe_2O_3, providing much better electrochemical activities. A thin film battery electrode comprising binder-free α-Fe_2O_3/MWCNT in combination with MWCNT (as the capacitive electrode) imparted an energy density of 50 Wh/kg at a power density of 1 kW/kg (Zhao et al. 2009). The much-improved charge storage characteristics can be attributed to the enhanced electronic conductivity of the binder-free battery and fast and uniform access to the insertion sites due to the thin film-based electrode, in the LIC.

- *Tin dioxide (SnO_2):* Tin dioxide is a cheap, environmentally benign, and abundantly available naturally occurring compound whose theoretical capacity stands at ~782 mA h g^{-1} (Li et al. 2018). This makes tin dioxide (SnO_2) a promising battery electrode material for lithium-ion capacitors. However, SnO_2 undergoes a large volume expansion during the process of charge/discharge, which can cause extensive polarization of the electrode material, aggregation of SnO_2

particles, and formation of unstable SEI films. This would drastically degrade the device performance with each cycle. Several methods are employed to address these issues. One method is to synthesize nanostructures of SnO_2 with different shapes, e.g., one-dimensional wire/rod/tube, two-dimensional nanosheets, and three-dimensional porous or hollow structures. The other method is to embed highly active SnO_2 nanoparticles inside a conductive carbide network (such as nanoporous carbon or graphene) to form a composite nanostructure. The results show that the carbide carrier not only can alleviate the volume expansion of the SnO_2 during charge and discharge but also provides a good conductive network. An LIC fabricated from tubular mesoporous carbon as the capacitive electrode and an SnO_2–C hybrid (ultrafine SnO_2 encapsulated in tubular mesoporous carbon) as the battery electrode showed a maximum energy density of 110 W h kg^{-1} and a maximum power density of 2960 W kg^{-1} with a capacitance retention of 80% of the initial value after 2000 cycles (Li et al. 2018).

- *Pseudocapacitive metal oxides (V_2O_5 and Nb_2O_5):* Although compounds like V_2O_5 and Nb_2O_5 are not explicitly battery-type materials (they belong to the group of extrinsic pseudocapacitors), there are still a good number of reports on their use as an electrode in lithium-ion capacitors. Now, technically, a hybrid metal-ion capacitor should have a battery-type electrode and another capacitive electrode. Furthermore, the combination of a pseudocapacitive material (positive electrode) and a capacitive material (negative electrode) is prevalent in asymmetric-type configuration, rather than in hybrid metal-ion systems. Having said that, these pseudocapacitive materials can produce battery-like behavior depending upon the redox environment. A summary of the report on both V_2O_5- and Nb_2O_5-based LICs has been provided in the below table (Table 3.2).

Table 3.2 Summary of the pseudocapacitive oxide (Nb_2O_5 and V_2O_5)-Based LIC Devices

Device configuration (anode// cathode)	Type	Voltage	Max energy (Wh/kg) @ power (W/kg)	Energy (Wh/kg) @ max power (W/kg)	Capacity retention	Reference
T-Nb_2O_5@C// MSP-20 activated	LIC	1–3.5 V	63@70	10@6500	75% over 1000 cycles	Lim et al. (2015)
Nb_2O_5– carbide-derived carbon// YP-50F	LIC	1–2.8 V	30@220	18@5000		Lai et al. (2017)
T-Nb_2O_5– graphene// activated carbon	LIC	0.8–3 V	47@393	15@18,000	93% over 2000 cycles	Kong et al. (2015)
Mesoporous Nb_2O_5–C// activated	LIC	1–3.5 V	74@120	20@12,137		Lim et al. (2014)
V_2O_5@CNT// activated carbon	LIC	0–2.7 V	40@210	6.9@6300	78% over 10,000 cycles	Chen et al. (2011)

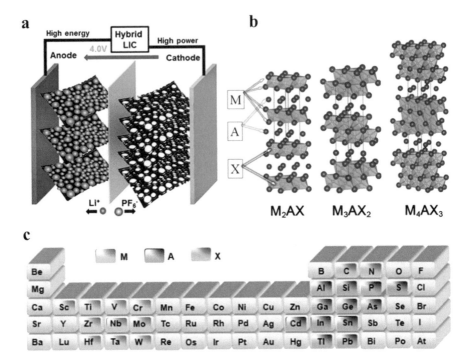

Fig. 3.9 (**a**) Energy-harvesting mechanism of LIC in the charging process. (**b**) Atomic structures of M_2AX, M_3AX_2, and M_4AX_3 phases. (**c**) The fragment of the element that makes up the MAX phase of the general composition $M_{n+1}AX_n$ in the periodic table. Reused with permission from Zhang et al. (2020a)

Interestingly, these metal oxides (especially Nb_2O_5) can yield impressive energy density and power density values. For instance, a maximum energy density of 74 Wh/kg was recorded for an LIC comprising carbon-coated mesoporous Nb_2O_5 and activated carbon as battery electrode and capacitive electrode, respectively. However, the power density is quite low, which might be ascribed to the low electronic conductivity of the Nb_2O_5 and its intrinsically sloping profile (which is a capacitive characteristic and fails to utilize the full energy spectrum), resulting in reduced specific energy.

Similarly, V_2O_5 goes through a bulk ion intercalation reaction and can readily accommodate ions like Li^+, Na^+, and K^+ in between the layered structures. In aqueous medium, the bi-layered phase of V_2O_5 accommodates Na^+ and K^+ efficiently in contrast to its orthorhombic counterpart. Furthermore, it has been observed that V_2O_5 can operate at comparatively higher working potential windows in organic electrolytes than aqueous media.

- *MXenes—layered metal carbides/carbonitrides*
  MXenes comprise a large number of layered transition metal carbides and carbonitrides (Fig. 3.9), which show bulk intercalation reactions ideal for metal-ion

capacitors (Couly et al. 2018; Zhang et al. 2020a). In aqueous media, these MXenes afford intercalation of a large number of cation species. For instance, reversible intercalation of Na^+, K^+, Mg^{2+}, and Al^{3+} cations have been observed in the case of multilayer bulk structure of exfoliated MXenes of the type $T_3C_2T_X$ (where "T" denotes the transition metals) (Come et al. 2012; Lukatskaya et al. 2013; Couly et al. 2018). However, despite the occurrence of bulk intercalation effect, MXenes produce capacitive cyclic voltammetry curves and sloping discharge profiles, which limits the specific energy values. Nevertheless, these compounds can produce extreme performances in combination with pristine or doped carbonaceous materials in a hybrid capacitor device. For example, three-dimensional TiC when coupled with nitrogen-doped porous carbon cathode yielded a maximum energy density of 101.5 Wh/kg and a maximum power density of 67.5 kW/kg. However, cyclic stability is one factor that needs to be addressed in the case of MXenes, and considering the fact that they belong to a completely new class of materials, extensive research activities are required in the area of structural and morphological optimizations to achieve efficient and stable performances.

- *Polyanionic compounds*

There are several additional titanium-based compounds, belonging to the pyrophosphate group, e.g., $LiTi_2(PO_4)_3$ and TiP_2O_7, which show appreciable lithium-ion insertion characteristics. However, the intercalation/deintercalation potential for both $LiTi_2(PO_4)_3$ and TiP_2O_7 is pretty close to ~2.5 V, which significantly reduces the working potential window for the fabricated LIC. For this reason, there are only a handful of reports on these polyanionic compounds as active electrode materials in lithium-ion capacitors. Both of these compounds possess robust 3D frameworks with multiple sites for adsorption of alkali ions (e.g., lithium ion). For example, nanosized ($\leq$100 nm) TiP_2O_7, carbon-coated $Li_3V_2(PO_4)_3$ (LVP-C), doped $LiTi_{1.5}Zr_{0.5}(PO_4)_3$ nanoparticles, and $TiNb_2O_7$ have been studied as battery-type electrodes in LICs. The compound $LiTi_2(PO_4)_3$ adopts a phase similar to the well-known NASICON-type (sodium superionic conductor) phase and yields high reversibility for lithium-ion intercalation and deintercalation. The three-dimensional crystal structure of $LiTi_2(PO_4)_3$ consists of corner-sharing PO_4 tetrahedra and TiO_6 octahedra (Aravindan et al. 2011; Peng et al. 2015; Satish et al. 2015). It is assumed that $LiTi_2(PO_4)_3$ could reversibly insert two lithium ions according to a two-phase mechanism between $LiTi_2(PO_4)_3$ and $Li_3Ti_2(PO_4)_3$. A pair of sharp redox peaks belonging to the Ti^{4+}/Ti^{3+} redox couple could be observed on the cyclic voltammogram of $LiTi_2(PO_4)_3$, cycled between 2.0 and 3.0 V. The shift between the generated redox peaks is considerably narrow, and the discharge plateau for $Li/LiCrTiO_4$, in a half-cell, was observed at ~2.4 V. The poor electronic conductivity of $LiTi_2(PO_4)_3$ is often tackled through carbon coating method, which significantly improves the power density value. When carbon-coated nanostructured $LiTi_2(PO_4)_3$ is combined with activated carbon in a typical LIC configuration, a maximum energy density of 14 W h/kg and a power density of 180 W/kg could be obtained. Similar performances were observed in the case of an LIC comprising the polyanionic TiP_2O_7 as the battery electrode and activated carbon as the capacitive electrode, where a maximum energy density of 13 W h/kg and power density of 371 W/kg were recorded.

- *Hydroxides (β-FeOOH)*

It has an analogous structure with the hollondite-type α-MnO_2, which is a promising cathode material for Li-ion batteries and can afford a theoretical capacity of ~300 mA h g^{-1}. β-FeOOH exhibits a tetragonal phase with 2 × 2 tunnel-like structure, which enables the facile insertion/extraction of Li ions (Amine et al. 1999; Aravindan et al. 2014). However, capacity fading is noted irrespective of the potential windows tested for 50 cycles. Li-HEC is fabricated using β-FeOOH with optimized mass loading of AC and cycled between 0 and 3 V. The Li-HEC delivered good capacitance retention properties up to 800 cycles and showed ~96% capacity retention at 10 °C rate. Further, the AC/β-FeOOH delivered an energy density of ~45 Wh kg^{-1}, which is twice that of the AC/AC symmetric non-aqueous system.

- *Silicon-based materials*

Silicon is known to have an excellent affinity toward lithium intercalation/deintercalation. In fact, in contrast to carbonaceous material (e.g., graphite) where six carbon atoms can take only a single Li$^+$ cation to form LiC_6, one silicon atom can afford to grab a whopping four lithium atoms at once. Additionally, silicon has a low lithiation potential of ~0.5 V (or less), and due to its strong lithium-ion intercalation property, the specific capacity can reach as high as ~3500 mA h g^{-1} (Yi et al. 2014). However, the extreme intercalation capacity is also a major drawback for silicon, as it can expand up to >300% of its original volume after being lithiated. This drastically affects the capacity of silicon-based battery-type electrodes. Furthermore, being an intrinsic semiconductor, the electronic conductivity of silicon is low. To address the volume expansion issue, silicon is often subject to nanostructurization processes in which the individual silicon nanoparticles do not have to experience volumetric strain as they can independently accommodate lithium ion. Also, smaller particle size reduces the ion diffusion path, and presence of a conductive additive-like carbon can solve the low conductivity issue. Highly porous silicon-based electrodes have also been reported, which did not show marked volume change during lithiation/delithiation. An LIC comprising boron-doped Si/SiO_2/C battery electrode and porous spherical carbon (PSC) capacitive electrode yielded an energy density of 128 Wh/kg at a power density of 1.229 kW/kg and an energy density of 89 Wh/kg at a power density of 9.7 kW/kg (Yi et al. 2014). The LIC retained 70% of the initial capacitance even after 6000 charge–discharge cycles.

3.4.1.2 Capacitive-Type Electrodes

The most common capacitive-type electrode used in lithium-ion capacitors is made from activated carbon. However, there are other carbonaceous materials that have the potential to be implemented as high capacity electrode materials for lithium-ion capacitors. Several such carbon-based structures have been listed below.

- *Carbonaceous materials*

Since an EDLC-type material is to be implemented as the counter electrode against the battery-type electrode in a lithium-ion capacitor, the most common choice is a carbon-based material with high specific surface area, good electronic conductivity, thermal and electrochemical stability, uniformly distributed pore channels, and good electrochemical accessibility. As we have discussed in the previous sections, a high specific surface area is an essential factor for these EDLC-type electrode materials to have appreciable capacitance, but the design and microstructural architecture of these carbonaceous materials can have significant impact on the charge storage properties of LICs. Take, for example, the case of activated carbon, which is widely regarded as an ideal capacitive electrode material for hybrid metal-ion capacitor devices. It has numerous randomly distributed pores throughout the three-dimensional structure, where the size of the pores can vary from micro range to macro range. The micropores can be as small as 0.3 μ and are hard to access if we talk about the ions present in the electrolytic media. Therefore, the capacity of the activated carbon electrode is greatly reduced. A strategic approach toward the fabrication of the carbon-based electrode is essential to bring uniformity in the distribution of the pores, well connected pore junctions to improve conductivity, and enhance the surface kinetics. Surface functionalization is the most used surface modification technique in the case of such carbon-based materials to improve surface adsorption and fast ion channeling by allowing the electrolytic content move through the uniformly distributed pores. Apart from activated carbon, other forms of carbon-based materials that are potential contenders of being used as the capacitive electrode materials in LICs include graphene and carbon nanotubes.

(a) *Activated carbon*: Natural fibers, carbohydrates, wood, bone chars, etc. are rich sources of these carbon-based materials. These sources undergo a carbonization process and another activation process (both at high temperatures) to produce activated carbon. Depending on the type of carbon source, temperature of the carbonization process, and the type of chemical environment in which the activation is carried out, we can have different forms of activated carbons. Low manufacturing cost, ease of synthesis, low coefficient of thermal expansion, and excellent thermal stability make activated carbon a strong candidate to be used as an electrode material in commercial devices. However, activated carbons have poor electronic conductivities, which is addressed by graphitization process or mixing certain high conducting materials such as carbon black (acetylene black).

Recently, many researchers have shifted their attention to various biomass wastes (e.g., corncobs, coconut coir, banana peels, egg white, etc.) due to the abundance and renewability of these raw materials. Biomass-derived AC materials with favorable physical and chemical properties, including good chemical stability, tunable microstructures, and surface functional groups, are recognized as the most promising electrode materials (Fig. 3.10).

(b) *Graphene*: Graphene has revolutionized the concept of two-dimensional materials in myriad of commercial applications and continues to do so with its

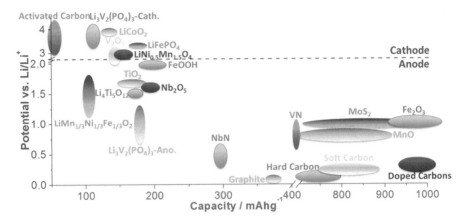

Fig. 3.10 Plot of the average working potential versus practical capacity of promising electrode materials for lithium-ion capacitors (LICs). Reused with permission from Ding et al. (2018)

unique 2D nanostructure having an enormous theoretical specific surface area as high as ~3000 m^2 g^{-1}. It also has excellent electrical conductivity, charge-transport mobility, and chemical stability. The micropores between the lamellae are favorable sites for the permeation of the electrolyte, and also it facilitates fast electron transport.

The fact that graphene has not yet been able to invoke sufficient commercial interest in the storage technologies relates to the difficulty associated with the synthesis of graphene. Therefore, the concept of graphene is still confined to the research laboratories only. However, various scalable methods have been explored to address this issue, but we are yet to see the final outcome in terms of large-scale commercialization. Currently, only the reduced form of graphite oxide, called as the reduced graphene oxide (rGO or RGO), is available at a large scale. But it does not fully mimic the physico-chemical properties of graphene and is affected largely by the various synthesis techniques as well as the reduction mechanisms involved.

(c) *Carbon nanotubes*: These are one-dimensional allotropes of carbon and exist in several different physical forms, i.e., single-walled carbon nanotube (SWCNT) and multiwalled carbon nanotube (MWCNT). These carbon structures are highly conducting in nature and share most of their properties with the two-dimensional graphene. In fact, these nanotubes are synthesized by the high temperature rolling of the graphene sheet to form the tubular structures. The finally obtained tubes can be open at both sides or stay closed, depending upon the chemical environment under which the synthesis has been carried out. Although CNTs are excellent materials of choice when it comes to high electronic conductivity and excellent mechanical and chemical stability, their high cost hinders large-scale commercialization. Also, they have low specific surface areas as compared to their two-dimensional counterpart, graphene. Nevertheless, many cost-effective methods have been proposed recently, which might pave the way for these wonderful carbon-based materials to enhance the storage technologies.

(d) *Other carbon-based capacitive electrodes*: Since graphene and carbon nanotubes are yet to be commercialized due to limited availability, therefore, activated carbon remains as the only viable option to be used as a capacitive electrode material. Besides, there are other forms of carbons such as porous graphitic carbon and highly ordered pyrolytic graphite, which have impressive specific surface areas and can replace activated carbon as high capacity EDLC-type electrodes in lithium-ion capacitors.

3.4.2 Electrode Materials for Sodium-Ion Capacitor

Although the standard redox potential of sodium is less negative than lithium and their ionic sizes do differ, they both show almost similar electrochemical activities. It can be safely assumed that the operational mechanism behind sodium-ion capacitor is almost equivalent to what we observe in the case of lithium-ion capacitors. Nevertheless, we can briefly discuss the types of electrode materials that are suitable for NICs.

3.4.2.1 Battery-Type Electrode Materials

Since the concept of sodium-ion capacitor has evolved around the concept of lithium-ion capacitor, therefore, much of the hybrid chemistry we would be discussing here would, of course, overlap with what we have already gone through in the lithium-ion capacitor section. Nevertheless, a brief account of the types of electrodes that have the potential to be implemented in the NIC devices has been provided.

- *Transition metal sulfides/selenides*

  Since carbon materials (such as hard carbon, graphite, etc.) have a low sodiation potential (~0.1 V), they tend to trigger dendritic growth at the electrode interface (Zhang et al. 2020b). Thus, there have been attempts to find appropriate battery-type materials for sodium-ion capacitor. Among them, transition metal sulfides/selenides (known as transition metal chalcogens) are gaining popularity recently. These compounds belong to the group of conversion- or alloy-type battery materials for sodium-ion capacitors. These are layered materials and have unique two-dimensional morphologies (considered the inorganic analogues of graphene) and possess diverse physico-chemical properties.

  MoS_2, a well-known 2-D layered compound belonging to this class of materials, has unique electronic and electrochemical features. However, agglomeration effect, subpar electronic conductivity (except for the 1 T metallic MoS_2), and volume expansion are some of the major drawbacks that hinder its application in sodium storage devices. However, these issues have recently been addressed through a unique interlayer expansion technique, which yielded ultrathin layers of MoS_2–carbon composite with an interlayer separation of ~1.02 nm (which is ideal for the insertion/deinsertion of sodium ion) (Wang et al. 2017).

Another example from this class of materials is tin sulfide (SnS_2 and SnS) (Zhao et al. 2019; Cui et al. 2019), which is also a layered two-dimensional compound and is considered as a potential sodium-ion storage material, due primarily to its high pseudocapacitance and low discharge plateau. Like MoS_2, it has a poor electronic conductivity too, which is addressed through composite formation using conductive carbon additives. Another compound belonging to this class is $MoSe_2$ (Ge et al. 2018; Zhao et al. 2018a, b), which has a narrower band gap (and slightly higher electronic conductivity) than that of MoS_2. Thus, it can be implemented as a potential electrode material for NICs. The techniques involved in this case (both synthesis methods and surface/structure modifications) require special set of synthetic routes, and their commercial scalability is yet to be considered. Considering the fact that the concept of NIC is still at a nascent stage if compared to the dominant LICs, these materials and associated synthesis procedures would, with time, be made feasible at a commercial scale through extensive research activities.

- *Titanium/Niobium compounds*

The redox potential of titanium falls within a range of 0.5–1.0 V, which prevents the plating (hence dendritic growth) of sodium metal. Thus, Ti-based compounds have much better cyclic stability in comparison to compounds like MoS_2. In fact, Ti^{4+}/Ti^{3+} redox-based anatase TiO_2 is probably the widest potential window electrode material for NICs, at present (Li et al. 2011; Bi et al. 2013; Liu et al. 2013). Being a metal oxide, it possesses a strong structural framework that is able to withstand the intercalation/deintercalation of Na ion without any marked volume expansion. However, poor electronic conductivity is still an issue. Furthermore, the diffusion rates of sodium ions are not as fast as lithium ions, thus nanostructurization of TiO_2-based electrode materials is often carried out to trigger faster ion diffusion. Besides, TiO_2 is often mixed with conductive carbon materials to improve the electronic conductivity. Another prospective material in this category is the lithium sodium titanate, which has a layered structure, for insertion/deinsertion of sodium ion. Since it has got sodium as a component, therefore, this material could avoid the additional processes such as pre-lithiation to compensate for the lost sodium ions. The first reported sodium titanate compound is $NaTiO_2$ and was explored for Na-ion intercalation. However, the instability of the compound led to the discovery of other forms of sodium titanate, e.g., $Na_2Ti_2O_4(OH)_2$ (Babu and Shaijumon 2017), $Na_2Ti_2O_{5-x}$ (Que et al. 2017), $Na_2Ti_3O_7$ (Yin et al. 2012), etc. to search for a stable structural geometry and appreciable electrochemical activity, suitable for NICs. In this case, the electronic conductivities of the compounds were improved by either mixing them together with conducting additive or growing them directly on a conducting substrate. Similar to the lithium-ion capacitors, sodium-ion capacitors also employ polyanionic compounds like NASICON-type $NaTi_2(PO_4)_3$ (Yang et al. 2018), monoclinic $Na_2Ti_9O_{19}$, etc. (Bhat et al. 2018), and two-dimensional MXenes such as $Ti_3C_2T_X$ layered structures for better ion diffusion and enhanced capacity. However, the stability at the electrode/electrolyte interface is one of the issues that needs to be addressed. Other forms of titanium-based compounds may include recently reported titanium oxynitrides, which have the general formula of TiO_XN_Y. Such compounds can be extracted from their respective metal-organic frameworks. N-doped porous carbon embedded with ultrasmall nanoparticles of titanium oxynitride produced a reversible capacity of 275 mA h g^{-1} at a mass normalized current of 50 mA g^{-1}.

Among the niobium compounds, Nb_2O_5 is the most prominent one. We have already mentioned Nb_2O_5 in the lithium-ion capacitor section for its lithium-ion storage characteristics. This compound exists in orthorhombic, pseudo-hexagonal, and amorphous phases, of which the orthorhombic phase has been found to have a superior Na-ion intercalating property, due primarily to its large lattice spacing of ~3.9 Å. However, like most of the metal oxides, Nb_2O_5 has poor electronic conductivity, which is addressed by forming its composites with highly conducting carbon materials. Furthermore, Nb_2O_5 has a low theoretical capacity (due to the single redox element, Nb), which limits the energy density of the NIC. Therefore, mixed titanium-niobium oxides, such as $Ti_2Nb_2O_9$ and $TiNb_2O_7$, have been identified and developed to be used as battery-type electrode materials for sodium-ion storage, because of multiple redox couples of Ti^{4+}/Ti^{3+} and Nb^{5+}/Nb^{4+} (Han et al. 2011; Han and Goodenough 2011; Li et al. 2015; Liao et al. 2020). Beyond that, other Nb-based mixed metal oxides were also designed and applied for sodium ion storage because they delivered excellent lithium storage performance.

- *Organic compounds*

   Organic materials are of low cost and easy to synthesize as compared to the inorganic compounds. They can provide a wide range of choices due to their flexibility and large window of compatibility. Recently, some of these organic-based compounds/composites have shown interesting metal-ion storage characteristics in order to create a green energy storage environment, which will cut off the carbon emission footprint significantly. A recent report shows an NIC comprising perylene-3,4,9,10-tetracarboxylic acid dianhydride (PTCD) as the battery material and polyaniline (PANI) as the capacitive material yielding a maximum energy density of 95 Wh kg^{-1} and a power density of 7 kW kg^{-1} (Thangavel et al. 2017). Another SIC was reported comprising disodium rhodizonate (DSRH) battery electrode and capacitive activated carbon electrode derived from cardamom shells (Thangavel et al. 2018). For organic-derived electrode materials and devices, the electrochemical performances including capacity, rate capability, and cycle stability can be improved through the following steps. First step involves forming composites with highly conducting carbon materials. Second step involves molecular engineering (Fig. 3.11). Manipulating the functional groups in organic molecules can effectively control the redox potentials. Third step is to modify the electrolytic content/mixture. The selection of a highly concentrated solvent/solute phase or solid electrolytes can reduce the solubility of organic materials to prolong the cycle lifetime. The next step is the polymerization technique in which design and tailoring of porous and conjugated structures can not only shorten the ionic transfer pathway to enhance the response capability to a large current density but also improve the cycle performance.

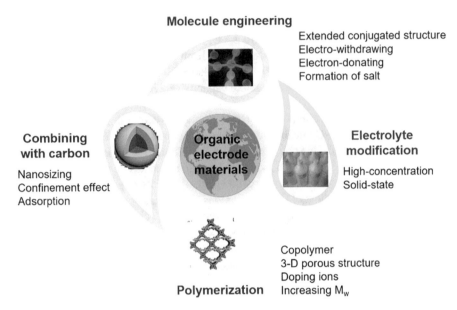

Molecule engineering

Extended conjugated structure
Electro-withdrawing
Electron-donating
Formation of salt

Combining with carbon

Nanosizing
Confinement effect
Adsorption

Organic electrode materials

Electrolyte modification

High-concentration
Solid-state

Copolymer
3-D porous structure
Doping ions
Increasing M_w

Polymerization

Fig. 3.11 Design strategies of organic materials for SICs. Reused with permission from Zhang et al. (2020b)

3.4.2.2 Capacitive Electrode Materials

For electrolyte consuming mechanism-based NICs, commercial grade activated carbon is the ideal choice as the capacitive electrode material (as evident from Fig. 3.12). In the electrolyte consuming mechanism, as we have already discussed in the sodium-ion capacitor section, the ion source is the electrolytic media, and the decrease in the sodium-ion concentration could drastically reduce the overall capacity of the hybrid device. This is the reason why, despite its high surface area and excellent adsorption capabilities, activated carbon possess relatively low capacitance, which limits the improvement in energy density of NICs. Several attempts have been made to design such carbon materials into various structures ranging from 1D to 3D, in order to expand the specific surface area and further to enhance the capacitance. Recently, biomass or organic compounds-derived carbon (Ding et al. 2016; Wang et al. 2019a), graphene (Wang et al. 2015; Zhang et al. 2018b), CNTs, etc. (Lota et al. 2007) have been investigated as potential capacitive electrode materials for NICs. However, the capacitance is still far less than battery materials, leading to a mass imbalance. One of the fundamental strategies to address this issue is by modifying the carbon framework with heteroatomic doping to enhance wettability and to provide pseudocapacitance. Doped carbon structures such as N- and B-doped carbon are quite capable of boosting the electrochemical performance because of the redox reaction of B- (or N-) containing groups or oxygenated groups around the B- (or N-) doping sites. However, different from B- and N-doping, S- or P-doping could lead to expanded interlayer spacing and structural defect sites due

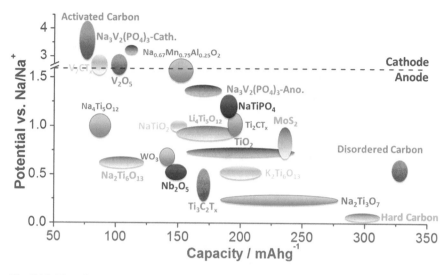

Fig. 3.12 Plot of the average working potential versus practical capacity of promising electrode materials for sodium-ion capacitors (NICs). Reused with permission from Ding et al. (2018)

to the larger atomic radii than C. Zhao et al. employed H_2O_2 to adjust the ratio of aromatic sulfide, sulfoxide, and sulfone groups in the S-doped carbon (Zhao et al. 2012). They postulated that the improved capacitance stemmed from polarized surface and reversible redox faradic reactions. In addition, P-doping in the carbon matrix effectively reduces the electrophilic oxygen species and enhances the oxidation stability, leading to an enhanced performance. Based on the independent property of B-, N-, S-, P-doped carbon, multiatomic co-doping, in the carbon structures, could achieve synergistic effects, which have been reported widely. However, the process of controlling the doping mechanism is way too complex, which is yet to be simplified. Besides these carbon or doped carbon materials, typical pseudocapacitive materials such as electroactive organic molecules (methylene blue, quinones, etc.), polypyrrole, and polythiophene are some of the promising candidates because of their high capacitance and fast kinetics.

3.4.3 Electrode Materials for Potassium-Ion Capacitors

Potassium-ion batteries have recently seen a surge of interest from the research community. The advantages with potassium-ion storage have already been discussed in the previous section. Considering the successful implementation of the lithium-ion capacitors in the light of LIBs, many researchers have already started showing interest in the extension of the current knowledge base surrounding KIBs for potassium-ion capacitors. Therefore, the concept of potassium-ion capacitor is still in its infancy. Nevertheless, we can have an overview on some of the potential electrode materials that can be used to fabricate high-performance KICs.

3.4.3.1 Battery-Type Electrodes

In potassium-ion batteries the battery-type electrode can be of pure potassium metal or other potassium-based compounds. Similar to the electrolyte consumption mechanism in the case of NICs, a KIC can also have an electrolyte medium that can act as a potassium-ion source. It is to be noted that since potassium has a low melting point, using it in pure metallic form is restricted to only temperature ranges that can be considered safe. At elevated temperatures, potassium intercalated compounds and/or other potassium-ion sources are generally employed. The battery-type electrodes in the case of KICs are almost similar to those found in the case of NICs, although potassium has a larger ionic size than sodium.

- *Titanium oxides*

  The most common inorganic compound employed for the insertion/deinsertion of metal ions is based on the Ti^{3+}/Ti^{4+} redox couple. Electrochemical insertion of lithium and sodium ions into titanium dioxide polymorphs (i.e., anatase, rutile) has been intensively studied in the past, with excellent outcomes. Alkali titanium oxides, such as the spinel $Li_4Ti_5O_{12}$ or the monoclinic $Na_2Ti_3O_7$, have the combined advantage of being low cost, ease of synthesis, and non-toxicity. Furthermore, protonated titanium dioxide has shown promising lithium insertion characteristics recently. Potassium analogues of lithium and sodium titanates, e.g., $K_2Ti_4O_9$, $K_2Ti_6O_{13}$, and $K_2Ti_8O_{17}$ (Dong et al. 2017), have also been investigated recently. Nanostructurization has been quite effective in enhancing the ion diffusion process for alkali metal ions. In fact, $K_2Ti_4O_9$ nanoribbons made from acid-leached Ti_3C_2 (MXene) obtained through a hydrothermal route exhibited higher capacity even at high current rates as well as longer cycling life. The inherent low electronic conductivity of these metal oxides is often tackled by mixing them with highly conducting carbon materials or through coating. However, for all these oxides, a low coulombic efficiency may be attributed to the possible formation of SEI or to a significant trapping of potassium ions in the layered structure of the electrode material.

- *Metal phosphides*

  Recently, a metal phosphide compound, Co_2P, has been prepared via a colloidal mesostructured method, for its possible application as the battery-type electrode material for KIC. The nanorod-shaped Co_2P, anchored to the surface of a reduced graphene oxide substrate, delivered a capacity of 374 mA h g^{-1} at a discharge current of 20 mA g^{-1}. The electrode showed an excellent rate capability of 141 mA h g^{-1} at 2 A g^{-1} (Wang et al. 2019b).

- *Polyanions (pyrophosphates)*

  Similar to LICs and NICs, potassium-ion capacitors employ polyanionic compounds in hybrid electrode devices. These polyanionic compounds possess opened frameworks that facilitate the ion diffusion process. Further improvements in the electrochemical performances of these compounds can be achieved through chemical substitution of either the transition element or the constituent ligand species. Since we have previously observed improved performances in the case of LICs and

NICs with these polyanionic compounds, it is thus expected that potassium-containing polyanionic compounds might also impart improved storage performances in KICs (Han et al. 2016; Hosaka et al. 2019). NASICON-type $KTi_2(PO_4)_3$ and hierarchical $Ca_{0.5}Ti_2(PO_4)_3$ compounds have been reported to have appreciable electrochemical performances (Zhang et al. 2018c).

- *MXenes*

The electrochemical performances of MXenes have been explored largely in the case of KIBs. This can be extended for KICs, because MXenes possess excellent layered properties and their interlayer separation can be tuned to accommodate fast insertion of alkali metal ions like K^+. MXenes are usually obtained by the leaching of the "A" layers in the pristine phase, $M_{n+1}AX_n$. It results in a chemical composition, $M_{n+1}X_n$, or $M_{n+1}X_nT_x$. Theoretical simulation of potassium insertion into Ti_3C_2 or O-terminated Ti_2CO_2 MXenes provides high capacities of 192 and 264 mA h g^{-1}, respectively (Er et al. 2014; Xie et al. 2014). $Ti_3C_2T_x$ (T = O, F, and/or OH) has also been evaluated as a potential electrode material, providing a first discharge capacity of 260 mA h g^{-1}. While the subsequent charge is still interesting (146 mA h g^{-1}), a continuous capacity fading is observed upon cycling (down to 45 mA h g^{-1} after 120 cycles). Three-dimensional porous alkaline Ti_3C_2 nanoribbons with an expanded interslab space can be easily obtained from Ti_3C_2 in aqueous KOH (Lian et al. 2017).

Although good discharge capacities are usually observed for the aforementioned titanium-based insertion-type anode materials, the subsequent irreversibility and rapid capacity fading dramatically lowers the overall performance. The insertion of a large amount of potassium ions in rigid crystalline structures does not seem straightforward, and only limited capacities are obtained. Moreover, capacity fading is often observed at elevated current densities, most notably due to the kinetic limitations. As the electrolyte is expected to influence the cycling performances of most of these electrode materials, therefore, revisiting these titanium-based materials with other electrolytic solvents/salts and formulations might lead to enhanced electrochemical properties.

- *Alloying- and conversion-type electrodes*

The problem with insertion/deinsertion type of electrodes is that their capacity degrades over time with repeated charge–discharge cycles. This might be due to several factors such as lattice straining, volume expansion, chemical instabilities due to strong interaction with the intercalating species, etc. This can be avoided by employing a set of alloying- or conversion-type electrode materials. The advantage with these materials is that they undergo a phase transition after the intercalation of the metal-ion species. The original compound is completely reshaped in order to accommodate the metal ion inside its interstices. This method ensures that the capacity of the device remains intact, even after thousands of charge–discharge cycles. Intermetallic compounds and conversion-type materials based on the p-block elements have shown interesting electrochemical activities recently. One of the promising alloying elements is antimony, because of its low working potential vs. K^+/K and its high theoretical capacity of 660 mA h g^{-1}, and forms an alloy of type K_3Sb (Gabaudan et al. 2019). Although there are a number of reports in this regard, still the reaction mechanism of Sb with K is not fully understood yet, due to the extensive amorphization of the material during cycling events, and requires strategic approach in order to identify the structural transformations that take place during cycling.

Bismuth is another element that can readily react with potassium to form K_3Bi, corresponding to a theoretical capacity of 385 mA h g^{-1}. Similar phenomenon has recently been observed in the case of an NaBi electrode, where the intercalation of Na ion triggered a conversion process to form Na_3Bi. Irrespective of the nature of the Bi electrode and of the electrolyte, a single potential plateau is observed around 0.35 V under galvanostatic conditions during the first discharge, suggesting a two-phase reaction. During the following sweeps, however, three flat plateaus can be detected between 0.5 and 1.3 V, suggesting three independent reaction steps (Gabaudan et al. 2019).

Tin is another well-known anode material in LIB and NIB, exhibiting a specific capacity of 991 and 847 mA h g^{-1} corresponding to the formation of $Li_{22}Sn_5$ and $Na_{15}Sn_4$, respectively. Moreover, tin is earth abundant, low cost, and non-toxic. In KIB, first principle DFT calculations suggested the formation of KSn as the most potassiated phase at the average potential of 0.5 V leading to a capacity of 226 mA h g^{-1} (Kim et al. 2018). The volume expansion resulting from the KSn formation is 180%, which is relatively low compared to formation of $Li_{22}Sn_5$ (257%) and $Na_{15}Sn_4$ (410%) in LIB and NIB, respectively (Ramireddy et al. 2017). Nevertheless, the potassiation of tin is also accompanied by many cracks inducing electrode pulverization and continuous electrolyte degradation with the exposition of fresh tin surfaces to electrolyte.

Other alloying-type materials such as Si, Ge, and Pb have been reported, but they have not been able to gather much popularity, as of now. These elements can react with only one potassium atom, leading to interesting theoretical capacities of 954, 369, and 129 mA h g^{-1}, respectively (Gabaudan et al. 2019). Regarding silicon, no reliable evidence was found for the electrochemical potassiation of an Si/graphene electrode in contrast with the predicted formation of KSi alloy. Phosphorus is also a promising electrode material by virtue of its high abundance and high theoretical capacity of ~2594 mA h g^{-1}. Potassium is expected to form K_3P with phosphorous. The K–P phase diagram presents the following range of alloys: K_3P, K_4P_3, KP, K_4P_6, K_3P_7, K_3P_{11}, and KP_{15} (Sangster 2010). However, phosphorus has an intrinsically low electronic conductivity and goes through a substantial volume expansion during alloying reactions. The use of intermetallic compounds and nano-structuration in porous carbon frameworks are possible solutions to overcome these drawbacks.

- *Organic compounds as electrodes*

Organic compounds are cheap and can be easily synthesized from a large pool of biomass reserves and their properties can be tuned according to the specificity of the application. There are a few compounds that have been explored in the context of KIBs. However, they might have the potential to be implemented as good battery-type electrodes in KICs as well. Some of the compounds belonging to this class of materials include dipotassium terephthalate ($K_2C_8H_4O_4$) (Deng et al. 2017), potassium 1,1′-biphenyl-4,4′-dicarboxylate (K_2BPDC), and potassium 4,4′-E-stilbenedicarboxylate (K_2SBDC) (Li et al. 2017). Besides, there are few oxocarbon salts [$M_2(CO)_n$, with n ranging from 4 to 6] like $K_2C_6O_6$, $K_3C_6O_6$, $K_4C_6O_6$, and $K_4C_5O_5$, which can be explored in the context of high-performance, all-organic KICs (Zhao et al. 2016).

3.4.3.2 Capacitive Electrode Materials

Similar to LICs and NICs, the capacitive electrode materials that a KIC uses include activated carbon, graphene (or its derivatives), and carbon nanotubes (both SWCNT and MWCNT). While commercial grade activated carbon is the frequently used capacitive electrode material for hybrid ion capacitors, both graphene and carbon nanotubes are gradually being considered as replacements. Graphene has excellent physico-chemical properties, and due to quantum confinement effect, it possesses the highest electron mobility for any two-dimensional structure. High specific surface area and robust electrochemical characteristics are some of the additional advantages possessed by graphene. Carbon nanotubes are technically the rolled-up versions of their two-dimensional counterpart, i.e., graphene. Since the quantum confinement effect gets more prominent with reduction in dimensionality, carbon nanotubes have better electronic conductivity than graphene, and it can form strong conducting nanotube networks that have a wide range of applications. Although graphene and carbon nanotube have huge advantages over the currently used activated carbon electrodes, high cost, and limited synthetic procedures have marred their commercial success. Nevertheless, this can be taken care of in the coming years with extensive research activities.

3.4.4 Electrode Materials for Calcium-Ion Capacitors

To compete with monovalent metal-ion capacitors, in terms of energy density, multivalent metal systems should be employed in their pure metallic form as one of the electrodes. This is an essential parameter for achieving highest possible energy density values from these multivalent metal-ion-based energy storage systems. One of the major issues with divalent metal-ion systems is that their strong charge densities would result in a higher desolvation energy, and they also tend to have very strong interactions with the intercalation cathode material, resulting in capacity degradation.

Furthermore, the electrochemical performances of these multivalent ion systems in their metallic form would strongly depend upon the type of SEI layer involved. Between Ca and Mg, the latter has been found to go under a plating/stripping process without resorting to any SEI layer formation. However, that has been achieved through rigorous electrochemical investigations on the compatibility of the Mg metal and the electrolyte (organic). Although Ca ion does form an SEI layer, the formed layer is not analogous to the one usually observed in the case of lithium-ion systems. The SEI layer formed by Ca metal is hardly permeable by the Ca^{2+} cation and results in performance degradation of the calcium-ion storage cell. The intercalation chemistry involving Ca ion is not fully understood yet, and there are plenty to be done before a calcium-ion-based hybrid storage technology is developed. Although not in a significant manner, few electrode materials have recently been reported showing good hybrid storage performances. They are detailed in the following sections.

3.4.4.1 Battery-Type Electrodes

These can be intercalation- or alloying-type electrode materials. Calcium is known to form alloy with other metals like aluminum, copper, and tin due to its biphasic property. Recently it has been found that an alloying/dealloying mechanism involving calcium and tin (Sn) could be used to facilitate the charge storage process by combining supercapacitor and dual ion battery techniques in a single system. An activated carbon-coated aluminum foil was taken as the capacitive electrode and calcium metal-coated tin foil was used as the bimetallic battery-type electrode. During charging the PF_6^- anions move toward the capacitive electrode, while Ca^{2+} ions move toward the tin electrode, where they trigger an alloying process with the tin foil by forming Ca_7Sn_6 (Wang et al. 2018). There are also varieties of hexacyanoferrates that have been reported to have shown Ca-ion storage in aqueous medium.

A similar attempt has been made in another work, where the alloying mechanism involving Ca and Sn has been exploited by taking manganese hexacyanoferrate as the intercalation medium and a tin foil as the alloying substrate (since calcium readily forms alloy with tin) (Wu et al. 2019).

Nickel hexacyanoferrate (NiHCF) has also been investigated for reversible insertion of Ca^{2+} along with various other divalent cations such as Sr^{2+} and Ba^{2+}. The aqueous electrolytes were prepared by dissolving appropriate metal nitrates in deionized water. As explained earlier, the electrode particle size, ionic conductivity, presence of zeolitic water, and open crystal structure of the active NiHCF were reasoned to help obtain the high rate capacities and high coulombic efficiencies. These observations certainly indicate the versatile nature of NiHCF with high power capabilities for cathode applications in multivalent aqueous electrolyte intercalation batteries (Gummow et al. 2018).

Copper hexacyanoferrate (CuHCF) was also investigated for insertion of several multivalent ions, which can readily be extended for Ca^{2+}, with good reversibility. Both ferricyanide vacancies and high degree screening of water molecules (from the crystal structure) in the intercalation of multivalent ions make CuHCF a unique electrode material for multivalent ion batteries (Gummow et al. 2018).

3.4.4.2   Capacitive Electrode Material

Till now, activated carbon has been employed as the only capacitive electrode material, and that too in a limited number of experiments. Since controlling the movement of Ca^{2+} and its subsequent adsorption on materials like activated carbon is a seemingly complex process, more research regarding the ion kinetics and electrode/electrolyte interaction is essential in finding potential capacitive-type electrodes for future Ca-ion storage (CIC).

3.4.5   Electrode Materials for Magnesium-Ion Capacitor

There are various categories of insertion (battery)-type electrode materials for Mg-ion storage systems, which can be extended for magnesium-ion capacitors (MICs) as well. Apart from the pure metallic magnesium, which can be used directly as an electrode against a capacitive electrode as the counter, here we list the various other electrode materials for MICs.

3.4.5.1   Battery-Type Electrode Materials

- *The chevrel phase:* The concept of intercalation/deintercalation can be derived from what we observed in the case of lithium, considering its similarity with Mg. However, despite the discovery of numerous potential battery-type materials, only the chevrel phase has shown excellent cyclability and relative high rate capability. The general formula of this phase is $M_xMo_6T_8$, where M = metal, e.g., Li, Mg, Cu, Zn, etc., and T = S, Se, and Te (Zhang and Ling 2016). The crystal structure of CPs consists of octahedral clusters of Mo (Mo_6) inside cubic anion frameworks as stacks of Mo_6T_8 blocks within a three-dimensional open structure. This Mo_6 cluster exhibits variable valence during the insertion and removal of Mg^{2+} ions, while the anionic framework is flexible with multidirectional paths for ionic diffusion. Chevrel-phase Mo_6T_8 can theoretically sustain up to four electrons exchanging, which corresponds to the insertion of two bivalent Mg^{2+} ions (Muldoon et al. 2014). This unusual structure provides a large number of close vacant sites and metallic Mo_6 clusters. The Mo_6 clusters are believed to be a critical factor to ensure high Mg^{2+} diffusivity in solid-state host because they can easily compensate the charge imbalance due to the introduction of bivalent Mg^{2+} ions.
- *Metal oxides:* These class of materials include Birnessite MnO_2, V_2O_5, MoO_3, and spinels (Mn_2O_4) (Muldoon et al. 2014; Zhang and Ling 2016) and have been investigated for the intercalation/deintercalation of metal ions since a long time. Among these oxide materials, MnO_2 has been studied extensively for its capability to accommodate monovalent as well as divalent metal cations. α-MnO_2 is a well-known compound used in zinc-based batteries and recently been studied for

Mg-ion secondary batteries as well. Metal oxides are, in general, structurally robust and have greater tolerance to sustain lattice strain during the intercalation/deintercalation. However, poor electronic conductivity remains as an issue.

- *Polyanion compounds:* We have already come across these kinds of compounds in the case of LICs, NICs, and KICs. These materials belong to the pyrophosphate group and have unique intercalation/deintercalation characteristics. Compared to the oxides, the induction effect of the polyanion groups moves the oxygen *p* orbital to deeper level, thus they generally provide better chemical stability of the cathodes. Research for polyanion compounds as Mg battery cathode mostly used 3*d* transition metal (V to Ni) as redox center to balance the charge neutrality when guest Mg is inserted or removed (Muldoon et al. 2014; Zhang and Ling 2016).

- *Transition metal chalcogenides:* These two-dimensional inorganic analogues of graphene are known to be excellent ion intercalation/deintercalation species. Since the research on the synthesis and characterization of transition metals and their compounds have extensively been carried out, there is a large base of materials in this category. Their tunable layered structures (through various techniques such as interlayer expansion techniques) and electrochemical signatures are noteworthy, making them excellent candidates as ion hosting materials. Some of these are also known for being conversion-type hosting compounds for ionic species with marked charge densities.

- *Sulfur:* Belonging to the chalcogen group, sulfur has been known for a long time as an excellent intercalation compound since the discovery of Li–S systems. Sulfur readily reacts with magnesium, and its theoretical capacity is $\sim$1671 mA h g^{-1} or $\sim$ 3459 mA h cm^{-3} (Zhang and Ling 2016). The combination of a magnesium anode and a sulfur cathode is of great interest because the theoretical energy density of Mg–S combination could reach over 4000 W h L^{-1}, which is approximately twice that of an Li-ion battery composed of a graphite anode and a cobalt oxide cathode. It indicates Mg–S battery can serve as a high energy storage system.

3.4.5.2 Capacitive Electrode Material

Although magnesium plating does not have to face the serious issue of dendritic growth, it suffers poorly due to the strong interaction with intercalation electrodes like graphite. Magnesium forms stage 1 graphite intercalation compounds, which makes graphite a poor choice for the host material. There are alternatives such as graphene, carbon nanotubes, etc., but more research needs to be done in this category to check the feasibility of these capacitive electrode materials to adsorb Mg^{2+}-ion species.

3.4.6   Electrode Materials for Zinc-Ion Capacitors

Since zinc has been extensively studied for its electrochemical properties, in different forms and cell configurations, zinc-based batteries consist of many variants of electrode materials. However, most of the zinc-based batteries belong to the category of primary cells, and only recently, the concept of zinc-ion rechargeable battery has been revisited by some of the researchers. One of the major advantages with zinc is its relatively higher standard redox potential (-0.76 V vs. SHE) as compared to lithium (-3.04 V vs. SHE), which means it can go through plating/stripping process well within the decomposition potential window of aqueous-based electrolytes. This is the reason why among all the discussed metal ions, zinc has the utmost potential to be used as a low-cost and environmentally friendly electrode material for metal-ion capacitors. Much of the chemistries involving zinc are restricted to non-rechargeable systems such as alkaline zinc batteries, zinc-air batteries, etc. No doubt these systems can help extrapolate the implementation of zinc in rechargeable systems as well, albeit the concerned system will need some time to settle down before it can go for full-scale commercialization. Nevertheless, it is expected that the success of zinc primary cells at both domestic and commercial level will surely be replicated in the field of secondary storage systems too.

Below is a list of materials that are currently being used in zinc-based primary as well as secondary cells:

3.4.6.1   Battery-Type Electrode Materials

Alkali and alkaline earth metals are highly reactive; therefore, they cannot be used in their metallic form (especially in aqueous media). Even so, they tend to form SEI layer in non-aqueous solvents to prevent corrosion and additional parasitic side reactions. Thus, rocking chair-type mechanism is employed in storage devices based on alkali and alkaline earth metals. Processes like lithiation, sodiation, magnessiation, etc. are additional steps that are followed in order to prepare a metal-ion source (anode), which will, in combination with the cathode and a suitable electrolytic medium, allow reversible intercalation/deintercalation during cyclic operations. This is not the case with zinc, which has excellent stability in aqueous media. Therefore, the battery material for zinc-ion system (whether it is a primary cell or secondary one) will be metallic zinc.

- *Intercalation facilitating compounds*

   Since zinc-ion battery runs on aqueous electrolytes, it has the option to pick a suitable electrode combination from a large pool of materials that are capable of triggering a reversible intercalation/deintercalation of zinc-ion. Following is a list of various cathode materials that are assumed to be ideal for zinc-ion capacitors.

(a) *γ-MnO$_2$:* The phase evolution, in a mesoporous γ-MnO$_2$ material, has been investigated during the intercalation of zinc ions via in situ synchrotron XANES and XRD (Alfaruqi et al. 2015). The results confirm that with the intercalation of zinc ions, the typical tunnel-type γ-MnO$_2$ gradually transforms to a spinel-type ZnMn$_2$O$_4$ involving two intermediate Mn(II) phases, i.e., tunnel-type γ-Zn$_x$MnO$_2$ and layered-type L-Zn$_y$MnO$_2$ (Xu and Wang 2019). With the extraction of zinc ions, most of these intermediate phases revert back to the γ-MnO$_2$ phase, indicating reversible phase transformation during electrochemical reaction. At a current density of 0.05 mA cm^{-2}, the mesoporous γ-MnO$_2$ cathode can deliver an initial discharge capacity of 285 mA h cm^{-2}.

(b) *α-MnO$_2$:* This compound is one of the most commonly used intercalation compounds in zinc-ion systems. α-MnO$_2$ has a large and stable 2 × 2 tunnel structure, which can accommodate the intercalated Zn^{2+} ions. The reversible capacity can be almost 100% maintained after 100 cycles at a discharge rate of 6 °C, showing reversible zinc storage and robust 2 × 2 tunnel structure during electrochemical reactions (Xu et al. 2009). However, the electrochemical reaction mechanism for MnO$_2$/Zn batteries is still under debate. Currently, three mechanisms are proposed in the aqueous α-MnO$_2$/Zn batteries: zinc intercalation/deintercalation, conversion reaction, and H$^+$ and Zn^{2+} co-insertion (Xu and Wang 2019).

(c) *ZnMn$_2$O$_4$:* This compound is the zinc analogue for the LiMn$_2$O$_4$, which was successfully implemented for lithium-ion systems. The zinc ions can be extracted from the structure (ZnMn$_2$O$_4$) in acid condition during the Mn^{3+} disproportionation reaction, while, with the increase in Ni content, the extraction of Zn^{2+} will decrease. The researching results appear that the spinel-structured materials are not quite suitable for the intercalation of zinc ions, while, with a certain content of defects (vacancies), the Zn-ion diffusion can be much easier due the lower electrostatic repulsion. Inspired by the understanding that the creation of defects in spinel materials can open additional pathways for the transportation of divalent ions. At high current of 500 mA/g, the ZnMn$_2$O$_4$ spinel carbon composite material can supply the specific capacity of 150 mA h g^{-1} for 500 cycles with retention of 94% (Zhang et al. 2016).

(d) *V$_2$O$_5$:* Built by sharing edges and corners of square pyramids chains, the vanadium pentoxide shows square pyramid-layered structure. More importantly, square pyramid layer of α-V$_2$O$_5$ can include water molecules or ions such as Na and Zn ions into the interlayers (Xu and Wang 2019), which may change the layered structure and significantly affect the discharge/charge processes and electrochemical performances of ZIBs. It is reported that Zn^{2+} cations can reversibly insert/extract through the layered structure of commercial V$_2$O$_5$ bulk. Moreover, the morphology of bulk V$_2$O$_5$ gradually develops into porous nanosheet structure after cycling, which is caused by the exfoliation during charging and discharging processes. As a result, the V$_2$O$_5$ porous nanosheets can deliver a very high reversible capacity of 372 mA h g^{-1} at a mass-normalized current of 5 A g^{-1} for over 4000 cycles (Yan et al. 2018).

(e) *Other V$_x$O$_y$ forms:* Constructed by distorted VO$_6$ octahedra by sharing corners and edges, vanadium dioxide has a special tunnel-like framework. The big tunnel-like framework can facilitate the transportation for the diffusion of inserted ions. In recent years, VO_2 (B) has been widely studied as potential electrode material in organic electrolytes with monovalent ions, while the investigation of VO_2 (B) in divalent/multivalent ion batteries is almost blank (Xu and Wang 2019). There are other vanadium oxide compounds in this class of materials that include V_6O_{13} and V_3O_7, which have recently showed appreciable zinc-ion storage characteristics.

(f) $M_xV_yO_z$ *(Vanadate compounds):* A vanadate compound of the type $H_2V_3O_8$ has been reported to exhibit zinc-ion intercalation. It delivered a high capacity of 423.8 mA h g^{-1} at 0.1 A g^{-1}, with capacity retention of ~94.3% for over 1000 cycles (He et al. 2017). The remarkable performance is owing to the layered structure of $H_2V_3O_8$ with large interlayer spacing, which facilitates the transportation of zinc ions due to lower resistance and enables the intercalation/deintercalation of zinc ions with a slight change in structure. Several other vanadate compounds such as $Li_xV_2O_5 \cdot nH_2O$, $Li_{1+x}V_3O_8$, $K_2V_6O_{16} \cdot 2.7H_2O$, and $Na_{0.33}V_2O_5$ (Xu and Wang 2019) have also been reported to have the potential to facilitate zinc-ion intercalation/deintercalation with appreciable efficiency.

(g) *Prussian blue materials:* These materials are intrinsically redox active and are used as redox additives in electrochemical cells. They have open framework structures, which contain zeolitic water and possess special physico-chemical properties. In a typical Prussian blue material $(KFe^{3+}Fe^{2+}(CN)_6)$, the Fe^{3+} ions and Fe^{2+} ions are octahedrally connected with the nitrogen ends and carbon ends of the CN^- groups, respectively (Xu and Wang 2019). One-half of the open sites in framework structure are occupied by the K^+ ions. As more K^+ ions get intercalated into the framework, some of the Fe^{3+} ions are reduced to Fe^{2+}. The color changes from blue to colorless. The yielded product is called Everitt's salt. With the K^+ ions extracted from Prussian blue, Fe^{2+} ions are oxidized. The color turns to yellow, and the product is called Prussian yellow.

(h) *Cu- and Zn-hexacyanoferrate (CuHCF and ZnHCF):* These are considered Prussian blue analogue compounds and have recently shown interesting zinc-ion intercalation chemistries. There are other forms of Prussian blue materials available, which may be investigated for their ion adsorbing capacity for metal-ion capacitors.

Several additional compounds such as MoO_2, Mo_2N, MoS_2 (Liu et al. 2017; Xu et al. 2018), etc. also have the potential to reversibly intercalate/deintercalate Zn^{2+}.

3.4.7 Electrode Materials for Aluminum-Ion Capacitor

The development of aluminum-ion battery (AIB) and subsequently AICs is still under development, and for the past few decades, researchers have not been able to control the battery chemistry of Al. The assumption that Al will give away Al^{3+} cation, which would replicate the lithium-ion system, with more energy density and efficiency, has not yet been realized in practice due to the strong polarization effect of AL^{3+} cation. Besides, the trivalent cation rapidly forms covalent bonds and have a strong affinity toward chloride species. This is the reason behind the formation of AlCl$_4^-$ and AlCl$_7^-$ ions (chloroaluminate ions) in the case of [EMIM-Cl]-AlCl$_3$, which is the only viable electrolyte found so far. Unlike zinc, aluminum cannot be used in its pure metallic form in aqueous media, as it will react with water in the presence of oxygen to form AL$_2$O$_3$ (and release hydrogen gas). This process, though not as prominent or vigorous as in the case of alkali and alkaline earth metals, still possesses challenges. Furthermore, the standard redox potential for Al^{3+}/Al^0 is −1.67 (vs. SHE), which means aluminum will trigger the decomposition of water before it can form the trivalent ions.

Nevertheless, the trivalent cation and exceptionally high volumetric capacity have drawn significant attention recently, and it is expected that a stable Al-ion chemistry would be formulated in the near future, which would boost the AICs.

3.4.7.1 Battery-Type Electrodes

If not Al^{3+}, the chloroaluminate ions (albeit heavy and large) have been reported to have similar migrating behavior as that of lithium ions in an LIC. Since the chloroaluminate anions are large and heavy, they need strong intercalation framework in order to reversibly migrate between the anode and cathode with the electrolyte facilitating the channeling of those anions. Pyrolytic graphite has been reported for rechargeable AIBs recently, showing good intercalating behavior. However, the coulombic efficiency was found to be 75%.

The inherent drawbacks of [EMIM-Cl]-AlCl$_3$ type of electrolyte include their toxicity, high cost, and undesired evolution of toxic gases during electrolyte preparation and while the cell is in operational mode, makes them less interesting for battery devices. Currently, this is the only effective electrolytic mixture for AIBs (and AICs). However, it has been observed that at mild pH values (alkaline pH), aluminum can show moderate electrochemical behavior. But corrosion and surface deformation remain the major bottlenecks.

However, there have been reports on several battery-type electrodes used in the case of aluminum-ion batteries, where an aqueous electrolyte was employed containing an aluminum-ion source. Here, aluminum was not directly taken as the anode, rather the electrolytic medium acted as the source for Al^{3+} ions. Some of those battery-type electrodes are as follows:

- *TiO$_2$:* The intercalation of the trivalent aluminum ion was first recorded in the case of TiO$_2$ electrode, in an aqueous electrolytic medium. The insertion/deinsertion process depended on the redox couple, Ti^{4+}/Ti^{3+}. It was also found that the insertion and extraction of Li$^+$ and Mg^{2+} were weaker for the same aqueous electrolyte as compared to Al^{3+} (Leisegang et al. 2019), which might be ascribed to the small ionic size of the Al^{3+} ion. These studies also indicated a predominant solid phase diffusion reaction for Al^{3+} insertion in the anatase TiO$_2$.

- *Copper Hexacyanoferrate (CuHCF):* The storage of trivalent Al ions in CuHCF has been investigated previously, in the context of AIBs, using a 0.5 M $Al_2(SO_4)_3$ aqueous electrolyte. The theoretical specific capacity of CuHCF is predicted to be around 58.9 mA h g^{-1}, based on the molecular weight of the compound (Leisegang et al. 2019). Since the fabrication of hybrid ion capacitors takes a lot of excerpts from the progress in the current battery technologies, much work is still to be done for the practical realization of AICs. Nevertheless, CuHCF, with its open framework and good intercalation properties, could become another potential cathode material for Al-ion capacitors.

3.4.7.2   Capacitive Electrode Materials

In a recent report, CNTs have been used to reversibly intercalate and adsorb the chloroaluminate ions by employing a pure metallic aluminum as the anode. The arrangement of the electrode itself mimics a hybrid storage device, though the ion species are not Al^{3+}. The appreciable performance of the CNT-based cathodes suggests that capacitive electrode materials like activated carbon, graphene, and other functionalized carbon structures could be ideal for AICs as well.

3.5   Electrolytes for Metal-Ion Capacitors

Electrolytes is a key component in electrochemical energy storage systems like batteries and supercapacitors. Therefore, the performance of a hybrid ion capacitor will be drastically affected by the nature and composition of the type of electrolyte medium used. However, since the discussion on any electrochemical set-up is always dominated by the concept of nature and feature of the electrode materials, the importance of the electrolytic content is often neglected, which should not be the case while discussing the performance metrics of an electrochemical device like batteries or supercapacitors. Most of the metal-ion capacitors, we have discussed till now, require a compatible non-aqueous electrolyte to safely operate. The only exception is the zinc that remains stable in aqueous media. In fact, zinc batteries perform best when aqueous electrolytes are employed. A generalized list of the minimal requirements of electrolytes should include the following: (1) it should possess good ionic conductivity, with high dielectric constant. This would facilitate fast ion transportation and minimize the rate of self-discharge; (2) the electrochemical stability window should be large enough to afford high energy density values and to prevent electrolyte degradation; (3) it should have a stable electrochemical interaction with other cell components like the separator, electrodes, and current collectors; (4) it should have an extreme tolerance level toward abusive cell operating conditions, and (5) it should be non-hazardous, non-toxic, and environmentally friendly.

Since there are diverse metal-ion systems involved in our discussion, let us briefly overlook the types of electrolytes suitable for each of these metal-ion systems to get a clear understanding of their overall strength as well as limitations.

3.5.1 Electrolytes for Lithium-Ion Capacitors

Depending on the type of solvent media used, lithium-ion systems can be divided into two broad categories, namely, aqueous lithium-ion capacitors and non-aqueous lithium-ion capacitors, where the former employs an aqueous-based electrolyte and the latter uses mostly organic or ionic liquid-based electrolytes.

(a) *Aqueous electrolytes:* These are basically aqueous solutions of lithium salts at desired molarities. A rocking chair type of ion migration occurs in this particular case, since lithium metal reacts violently with water and oxygen. Compared with organic electrolytes, aqueous electrolytes have smaller solvated ions, higher ionic concentration, and higher ionic conductivity. Besides, aqueous electrolytes are easy to synthesize, ultra-low-cost, non-flammable, ultra-safe, and they have very low viscosity as well. Higher ionic conductivity means aqueous electrolytes would allow faster charge transfer (hence more power densities) than non-aqueous electrolytes (where ionic conductivity is usually poor). Moreover, aqueous electrolytes can be easily produced on a large scale and used without any preconditioning or precautionary steps, while hygroscopic organic ones usually need strict conditions and complicated processes to avoid any contact with ambient moisture. The choice of a suitable electrolyte lies in the sizes and radii of both hydrated cations and anions, and the ionic conductivity. Commonly used aqueous electrolytes comprise an aqueous base in which various salts such as $LiOH$, $LiCl$, and Li_2SO_4 are dissolved in desired molarities to achieve alkaline, acidic, and neutral solutions, respectively. One of the major shortcomings with aqueous electrolytes is their narrow electrochemical stability window, which may not exceed ~1.23 V under normal condition (1.23 V vs. SHE is the standard potential at which water dissociation is triggered). However, the water dissociation potential is dependent on the salt (ionic) concentration and the type of electrodes used in the electrochemical cell. So, by strategically tuning the controlling parameters, the electrochemical stability window can be enhanced to some extent. Still, aqueous electrolytes cannot match the working potential window in case of a non-aqueous, organic electrolyte. Neutral electrolytes are preferable, as low and/or high pH electrolytes can be corrosive. In practice, mild pH aqueous electrolytes are used in order to take advantage of the acidic or alkaline environment and to suppress the corrosive side reactions between the electrolyte and the cell components. Recently, redox-active electrolytes have attracted more attention due to the redox active components, which can contribute to the faradic charge-transfer process, thereby reducing the charge-transfer resistance of the cell. The commonly used electrochemi-

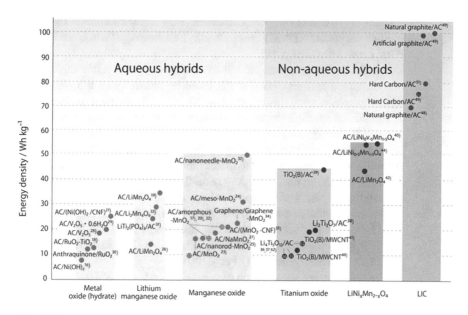

Fig. 3.13 Proposed hybrid supercapacitor systems in aqueous and non-aqueous media. There is difficulty in expressing the representative value for the energy density (based on weights of active electrode materials) of each hybrid system as it depends on the power. Reused with permission from Naoi et al. (2012)

cally redox-active compounds are hydroquinone, *m*-phenylenediamine, ferro-cyanides (or ferricyanides), potassium iodide, lignosulfonates, etc.

(b) *Non-aqueous electrolytes:* Most of the lithium-ion systems do use non-aqueous electrolytes; the reason is obvious that it does not react with lithium, and even if it does, then the formation of SEI layer takes care of it inside the cell. The major advantage that the non-aqueous, organic electrolytes have is their much larger working potential window as compared to the aqueous counterparts (Fig. 3.13). The composition of an electrolyte is critical, and it has several key components such as:

- *Solvent*—An organic solvent is basically a liquid matrix that provides a highly mobile ion matrix when a suitable salt is dissolved in it. An ideal solvent should have few desired characteristics; it should have a high solubility constant. Although organic solvents generally do not have high solubility indices, yet they can specifically dissolve few salts to achieve satisfying ion concentrations. Furthermore, the solvent should have a low viscosity, which would allow fast movement of ions. In addition, it should be least reactive toward rest of the cell components. Among organic solvents, only those with polar groups such as sulfonyl (S=O), nitrile (C ≡ N), carbonyl (C=O), and ether (-O-) functional groups can dissolve sufficient amounts of lithium salt (Li et al. 2018). The types of organic solvents for LICs include propylene carbonate (PC), ethers, ethylene carbonate (EC), linear carbonates including dimethyl carbonate (DMC), propylmethyl carbonate (PMC), and ethylmethyl carbonate (EMC), etc.

- *Lithium Salts*—The aforementioned solvents require a suitable lithium salt to be dissolved in them, so that the ionic conductivity of the electrolyte would improve. A good lithium salt has the following characteristics: (1) the salt should have a high degree of polarity, which would help it in getting dissolved in the organic solvent; (2) the negative ion of the salt should have a weak affinity toward lithium cation; (3) the cation–anion asymmetry of the salt should not be large enough to affect the capacity of the capacitor device; (4) the salt should be non-hazardous, non-flammable, least toxic, safe, environmentally friendly, and cheap.

- Lithium salts can be divided into inorganic lithium salts and organic lithium salts. $LiPF_6$, $LiAsF_6$, $LiBF_4$, and $LiClO_4$ are typical inorganic lithium salts. Organic lithium salts include $LiB(C_2O_4)_2$, $LiP(C_6H_4O_2)_3$, Et_4NBF_4, and lithium trifluoromethanesulfonate (Li et al. 2018).

- *Additives*—These are special kind of substances, when added to the solvent–salt mixture could drastically enhance few characteristics of the final electrolytic content. Basically, adding these substance/compounds is one of the most economical and effective ways to improve the performance of LICs. In general, the use of additives can significantly improve some properties of the LICs, such as the ionic conductivity of the electrolyte, the matching performance of the positive and negative electrodes, the capacity of the LICs, the cycle efficiency, the cycle life, the reversible capacity, and the safety performance, without substantially increasing the cost (the proportion of additive addition does not exceed 5% of the electrolyte) (Li et al. 2018). The additive element improves the overall ionic conduction in the electrolyte mixture. It also can facilitate the formation of smooth and effective SEI, helping fast ion accumulation. It should stabilize the working potential window by suppressing the parasitic reactions and bring thermal stability to the electrolyte.

3.5.2   Electrolytes for Sodium-Ion Capacitor

Sodium is also reactive like lithium, and even if lithium has been investigated with various aqueous salt solutions as electrolytes, sodium-ion capacitors have limited pool of electrolytes, and all of them are mostly organic or aprotic. However, there are few reports on aqueous-based electrolytes. Much work is needed to be carried out toward high-performance electrolytes along with the search for suitable electrode materials. All the characteristic requirements for an electrolyte for LIC also apply to the electrolytes that are going to be discussed here.

(a) *Aqueous electrolytes:* As we have already discussed in the previous section (for LICs), except for their narrow electrochemical stability window, aqueous electrolytes possess huge advantages over their non-aqueous and/or organic counterparts. The commonly used salts for aqueous NICs are NaCl, Na_2SO_4, $NaNO_3$, NaOH, etc., which can provide a sodium-rich environment. These salts are easily soluble in aqueous base and provide excellent ionic conductivities during device operation. Apart from these salts, sodium perchlorate ($NaOCl_4$) also acts as a sodium-ion source in aqueous solutions. The advantage with $NaOCl_4$ is that it has a very high solubility in water, and therefore, excess amount of sodium perchlorate is often dissolved in aqueous base to have a high concentration ionic liquid-type system. This prevents the electrolyte to decompose at potentials way beyond the thermodynamic degradation potential of 1.23 V (for water).

(b) *Non-aqueous/organic electrolytes:* The organic electrolytes mentioned in the case of LIC will also be applied in the case of NICs. Same set of solvent media, i.e., propylene carbonate (PC), ethylene carbonate (EC), and few ether-based solvents like diethylene glycol dimethyl ether (DEGDME), tetraethylene glycol dimethyl ether (TEGDME), and dimethoxyethane (DME) may be used in the case of NICs (Thangavel et al. 2016; Zhu et al. 2017; Le et al. 2017). Ether-based solvents could suppress electrolyte decomposition by forming a permeable solid–electrolyte interface (SEI) film on the graphite surface. In contrast, thick SEI films were formed on the graphite surface in the carbonate-based solvents, which block sodium-ions transport. The good electrochemical performance of the graphite in ether-based electrolytes develops new avenues for the development of sodium-ion energy storage. It is foreseen that ether-based solvents will emerge as a strong candidate in the near future. Sodium analogues of $LiPF_6$, $LiAsF_6$, $LiBF_4$, and $LiClO_4$ (e.g., $NaPF_6$, $NaAsF_6$, $NaBF_4$, and $NaClO_4$) salts can be used in the case of NICs.

Few ionic liquids have also been proposed, but currently only limited to the sodium-ion battery systems. Furthermore, polymer-based electrolytes like gel-polymers have also been found in the form of poly(vinylidene fluoride) (PVDF), polyacrylonitrile (PAN), and poly(vinylidene fluoride-hexafluoropropylene) (PVDF-HFP).

3.5.3 Electrolytes for Potassium-Ion Capacitors

Similar to LIC and NIC, the electrolytes for KICs can be both aqueous and non-aqueous/organic. However, respective salts and their solubility in the solvent media would determine the type of electrolytic performance in each case. Here also, for KICs, carbonate- and ether-based electrolytes are used. For ionic liquid case, water-in-salt concept is applicable for all the three systems, where a concentrated salt solution will mimic an ionic liquid-type environment. For aqueous electrolytes, salts like KOH, K_sSO_4, K_2NO_3, KCl, etc. can be used. Potassium perchlorate ($KOCl_4$) may also be used as in the case of LICs and NICs.

Similarly, for organic solvents, potassium analogues of $LiPF_6$, $LiBF_4$, and $LiClO_4$ (e.g., KPF_6, KBF_4, and $KClO_4$) salts can be dissolved to achieve good ionic conductivity. It is to be noted that potassium is more reactive than lithium, and therefore, it requires careful selection of the cell components, and a compatible electrode/electrolyte/separator configuration is essential for safe and efficient potassium-ion capacitors.

3.5.4 Electrolytes for Calcium-Ion Capacitors

Pioneering works on calcium-ion batteries were done by dissolving calcium perchlorate, $Ca(OCl_4)_2$, in three different solvent media, namely, γ-butyrolactone (BL), propylene carbonate (PC), and acetonitrile (ACN), and cycling the calcium metal anode (Muldoon et al. 2014). However, the dissolution and recovery of calcium were not observed due to the fact that Ca^{2+} is unable to penetrate through the SEI layer, unlike lithium-ion capacitor, where SEI facilitates smooth and dendrite-free dissolution and deposition of lithium.

However, at elevated temperatures (75–100 °C), the deposition of calcium has been observed. This suggests that at normal temperatures, the SEI formed on the anode (calcium metal) in the calcium-ion system does not allow the calcium ion to pass through. Although at elevated temperatures, it does show cyclic behavior; for practical application, this method does not seem to be too feasible. Only recently, it has been found that calcium in combination with an alloying metal-like tin (Sn) would go through the dissolution/deposition process. Recent reports suggest that the calcium deposition and stripping is possible if calcium is allowed to go through an alloying conversion process with metals like Sn, Al, or Cu, in a non-aqueous electrolytic environment.

Calcium-ion rechargeable battery has also been reported for aqueous electrolytic environment. Instead of taking a calcium metal as anode, a calciated copper hexacyanoferrate (CuHCF) was taken as the anode and cycled against a PANI-coated carbon cloth (PANI/CC). A 2.5 M aqueous solution of $Ca(NO_3)_2$ was taken as the electrolyte in this case (Muldoon et al. 2014).

Further attempts are being made to tune the electrolytic contents, by taking note of the performance enhancement at elevated temperatures, and also the possibility of aqueous calcium-ion capacitor (although the potential window is small in aqueous media), so that the concept of calcium-ion capacitor and calcium-ion battery could be realized in practice.

3.5.5 Electrolytes for Magnesium-Ion Capacitors

The case with magnesium-ion system, when it comes to compatible electrolytes and salts, is quite different despite its striking similarity with lithium-ion system. Both aqueous and non-aqueous Mg-ion systems have been reported. They have been categorized as follows:

(a) *Aqueous electrolyte:* Similar to calcium, the charge storage characteristics of magnesium have been investigated in both aqueous and organic electrolytes. For aqueous electrolyte case, a magnesium salt of the type $MgSO_4$ could serve as the Mg^{2+} ion source. Other forms of magnesium salt, e.g., $Mg(NO_3)_2$, $MgCl_2$, $Mg(OH)_2$, etc. may be used in aqueous base for aqueous magnesium-ion capacitors.

(b) *Non-aqueous electrolytes:* Magnesium reacts with most of the solvent bases having carbonate, sulfoxide, or nitrile groups, except for ether. Thus, ethereal solvents are preferred for Mg-ion systems. The unique electrochemistry of Mg prohibits its reversible deposition in aprotic solvents containing ionic salts such as magnesium bis(trifluoromethane)sulfonimide ($Mg(TFSI)_2$) or magnesium perchlorate $Mg(ClO_4)_2$. Furthermore, the reduction of electrolytic components results in the formation of a surface layer (SEI formation), which is apparently nonconductive with respect to both electrons and magnesium ions and thus inhibits Mg deposition.

Replacing Mg-based Grignards with $Mg(BH_4)_2$ produced effective results. Furthermore, a combination of $Mg(BH_4)_2$ with glyme was found to be superior to tetrahydrofuran (THF). The deposition of magnesium was further enhanced by adding $LiBH_4$ to the electrolyte mixture. The electrolyte, $Mg(BH_4)_2/LiBH_4/diglyme$ (Muldoon et al. 2014), showed excellent compatibility with magnesium metal.

(c) *Polymeric gel electrolytes:* These are viscous or freestanding gels comprising a polymer matrix, an alkali salt, a plasticizer (usually in the form of an organic solvent), and various additives. Thin films of these polymers can be formed to compensate for their lower conductivity values as compared to the liquid electrolytes. They have advantages over their solid-state battery counterparts because they are soft and conform to solid surfaces, resulting in better interfacial contact with various electrode materials. Use of gel polymer electrolytes also avoids the problem of cell leakage. Examples of such polymeric gels include poly(ethylene oxide) (PEO), EtMgBr−PEG−THF (ethyl magnesium bromide-polyethylene glycol-tetrahydrofuran), etc.

(d) *Ionic liquids:* Due to its high negative standard potential, plating/dissolution of magnesium can only occur in aprotic solutions. However, electrolytes composed of existing commercial magnesium salts in aprotic solvents with high dielectric constants such as carbonates, esters, or acetonitrile have been reported to form dense passivating layers, which are impermeable to the magnesium ion and block any deposition/dissolution. Successful plating/stripping has been observed in the case of magnesium organohaloaluminates or magnesium organoborate salts, which contain the $(Mg_2(\mu\text{-Cl})_3 \cdot 6THF)^+$ cation dissolved in etheric solvents. Ethers have a low dielectric constant, typically below 10, which restricts the solubility of magnesium salts (<0.5 M) as well as their ionic dissociation. It is desirable for a battery electrolyte to contain the electroactive salt in high concentrations (state-of-the-art Li-ion electrolytes contain the $LiPF_6$ salt at concentrations above 1.0 M) such that high conductivities and rapid charge/discharge rates can be achieved. In order to increase the solubility of currently available magnesium salts and diminish the volatile and flammable nature of ether-based electrolytes, room-temperature ionic liquids (RTILs) have been considered as potential solvents for a variety of magnesium salts (Muldoon et al. 2014). A typical RTIL is 1-n-butyl-3-methylimidazolium tetrafluoroborate (BMImBF$_4$).

3.5.6 Electrolytes for Zinc-Ion Capacitors

Zinc-based chemistries mostly take place inside aqueous environments, i.e., a simple solution mixture of water and various salts of zinc, e.g., ZnCl$_2$, Zn(NO$_3$)$_2$, ZnSO$_4$, Zn(OCl$_4$)$_2$, etc. However, there are few reports in which zinc has been treated with a non-aqueous electrolyte. Both types of electrolytic media are discussed below:

(a) *Non-aqueous electrolyte*: Recent trends in non-aqueous zinc-ion rechargeable systems commonly use a zinc salt (Zn(TFSI)$_2$, Zn(OTf)$_2$) dissolved in acetonitrile solvent. Prior to this, zinc non-aqueous systems used room-temperature ionic liquid electrolytes (e.g., trifluoromethanesulfonic imide) (Dong et al. 2019), owing to their very low vapor pressure, extremely high chemical/physical stability, and high ionic transportation. However, the discharge capacities and cycling lifetime of batteries based on these electrolytes are not satisfactory. As such, the acetonitrile–Zn(CF$_3$SO$_3$)$_2$ electrolyte shows a high anodic stability and a relatively low overpotential, leading to an excellent coulombic efficiency of the Zn anode.

(b) *Aqueous electrolytes:* These electrolytes are dominant when we talk about zinc batteries, albeit the current rechargeable systems are preferring non-aqueous electrolytes (for obvious reasons). Zinc can go under redox reaction in neutral, alkaline as well as acidic aqueous electrolytes (Fig. 3.14). However, the pH change should be mild; otherwise, there will be detrimental effect on the Zn metal anode.

- In mildly alkaline pH conditions (e.g., in 6 M aqueous KOH solution), Zn attaches itself to the hydroxyl groups and forms zincate ions that migrate far from the electrodes and finally settle down as ZnO, setting a course of irreversible metal loss. That is why alkaline cells employing zinc are primary cells and cannot be recharged.

- At mildly acidic pH (pH = 4–6), the pH window allows the Zn^{2+} ion to remain in its ionic form by suppressing electro-active reactions (in alkaline conditions). By employing salts such as ZnSO$_4$ or Zn(CF$_3$SO$_3$)$_2$ that offer mildly acidic aqueous solutions, reversible Zn plating/stripping becomes possible to a certain extent. Thus, recharging the battery becomes much easier. However, lower pH values can trigger hydrogen evolution reaction at potentials relatively higher than their standard values. However, this is tackled by the kinetic overpotential of zinc and by reducing the C-rate for the battery system.

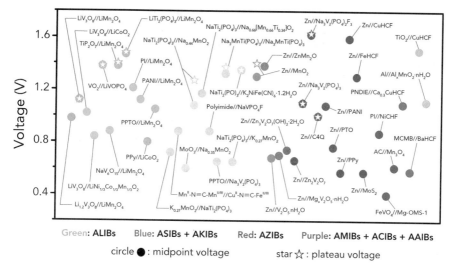

Fig. 3.14 Voltage comparison among various ARMB systems. Solid circles represent midpoint voltages, and hollow stars at the same x-axis position represent corresponding plateau voltages (the voltage of the first discharge plateau is adopted if a battery yields more than one plateau). Some sodium–potassium hybrid batteries are also included. Reused with permission from Liu et al. (2020)

3.5.7 Electrolytes for Aluminum-Ion Capacitor

The most commonly used electrolyte for rechargeable Al-ion batteries is [EMIM-Cl]-AlCl$_3$, as we have already discussed in the above section. However, there are reports where aqueous solutions of aluminum salts (e.g., Al$_2$(SO$_4$)$_3$) have been employed in Al-ion batteries.

Aqueous electrolytes based on AlCl$_3$ or Al$_2$(SO$_4$)$_3$ salts in water have been widely used for AIBs for low-cost, simple operation, and environmental friendliness. However, in aqueous electrolytes, a passive oxide layer (Al$_2$O$_3$) will form and the reversible deposition of Al^{3+} is hindered by a competing intrinsic hydrogen evolution reaction, leading to a low efficiency of the Al anode.

To overcome these limitations, non-aqueous electrolytes have been applied in AIBs. Initially, molten salts such as the binary NaCl-AlCl$_3$ system or the ternary KCl-NaCl-AlCl$_3$ system were considered as possible electrolytes for AIBs. In these molten salt systems, bare Al^{3+} does not exist, but in the forms of AlCl$_4^-$ and Al$_2$Cl$_7^-$. However, these molten salt electrolytes have a high melting point, which limits the practical applications of AIBs. Therefore, ionic liquids with high ionic conductivity, low volatility, and high electrochemical/chemical stability were then investigated as room-temperature electrolytes for AIB systems. The most widespread ionic liquid electrolytes for AIBs are AlCl$_3$ in imidazolium-based ionic liquids, in which imidazolium cations are typically 1-butyl-3-methylimidazolium or 1-ethyl-3-methyl-imidazolium (EMIM), etc.

There is no one-size-fit-all option for electrical energy storage, and therefore, exploration of non-lithium-based energy storage is essential in order to stabilize the stress on currently available primary as well secondary storage options. The promising aspects of the multivalent metal-ion capacitors are interesting, since a low-cost and environmentally friendly storage technology will be multi-fold effective than the current lithium-ion system in addressing the economic and geopolitical constraints associated with lithium. Nevertheless, simply drawing a comparison in terms of theoretical volume capacity or specific capacity would not be sufficient, since the performance of a metal-ion capacitor depends upon the coupling between the two electrodes, electrochemical and thermal stability of the electrolytic medium, and various other parameters such as ion diffusion and mobility, charge-transfer characteristics, and so on.

3.6   Measurement Techniques

Measuring the performance of a standard supercapacitor device involves a two-step process, where the first step is the material characterization and the second step is the device characterization. While the material characterization is done in a three-electrode configuration, device characterization is taken care by a typical two-electrode electrochemical cell. The details of the two systems have been provided in the following subsection.

3.6.1   Three- and Two-Electrode Configurations

Although a seemingly simple technique, three-electrode electrochemical analyses can provide critical information such as the electrochemically stable working potential window and reduction–oxidation processes through various potentiostatic, potentiodynamic, and galvanostatic techniques. A typical three-electrode system consists of a working electrode, a counter electrode, and a reference electrode. All the potential and current values are applied between the working electrode and the reference electrode (which serves as the reference point for all the recorded potential values). The counter electrode is, in general, made of a highly conductive and non-reactive element such as gold and platinum. The purpose of the induction of a counter electrode is to provide a path for the electrons to travel through the external circuit, so that charge equilibrium is maintained.

A two-electrode set-up can be thought of as a modified form of the three-electrode configuration, where no additional/auxiliary electrodes, e.g., reference or counter, are present. It consists of two working electrodes on each side, and the potential values are recorded by treating any of the working electrode as the reference point. As discussed before, the working electrodes can be of similar (symmetric) or different (asymmetric) electrochemical behaviors. Unlike the three-electrode system, which is actually a half-cell configuration comprising a single working electrode, two-electrode configuration provides a practical prototype/device architecture that can be used to evaluate essential parameters such as specific energy, specific power, long cyclic stability, etc.

Supercapacitance measurements, using both the configurations, i.e., three- and two-electrode configurations, are the most commonly implemented practice among researchers, as they reveal critical information regarding the electrode material and/or device features. Measurements using these conventional electrochemical configurations have been standardized by the International Electrochemical Commission (IEC). Three-electrode configurations are electrochemical half-cells with very high sensitivities, where even a single electron transfer process could be traced with the help of a combination of working-reference electrodes working in tandem, recording every minute of the current–potential fluctuations occurring inside the system.

Three-electrode configurations are not suited for carrying out full-scale analysis of a purely capacitive storage device. First of all, the half-cell in a three-electrode system is assumed to be an infinite system, where the other half of the complete cell is considered to be at a fairly large distance (large enough to let the half-cell run almost independently). Second, during anodic and cathodic scan, the charge accumulation occurs near the surface of the working electrode only, and since it is the only electrode present in the system, the rate of accumulation of charge can be as large as two- to four-fold greater than a two-electrode configuration, where the two electrodes maintain a mass–charge balance by sharing the charge as per their individual capacities. Thus, in most cases, discrepancies have been observed, when attempts are made to draw a comparison between the results obtained from both two- and three-electrode configurations. Apparently, the performance of a given material (in a three-electrode half-cell) exceeds well beyond the performance of the same material in a two-electrode set-up. Third, there must be a finite separation between the two electrodes of a capacitor. Here, the electrodes are assumed to be far apart from each other (because of the non-linear nature of the system), therefore it is not feasible to draw comparison between a three-electrode material characterization technique with a two-electrode device.

There are several reports in which the three-electrode system is modified in such a way that the counter electrode (which usually does not take part in any electrochemical activities inside the cell) is replaced by another electrode similar to the working electrode, so that the whole arrangement would mimic a two-electrode set-up. As we have already discussed, this arrangement does not replicate a capacitor device, since in a capacitor, the electrodes are meant to be placed within a finite separation. Nevertheless, this kind of arrangement holds good for redox systems, where specifically designed materials or electrodes are placed together inside a working electrolytic medium and allowed to exchange ions (inside the electrolytic medium) and electrons (through an external circuit). This kind of arrangement is semi-infinite in nature.

3.6.2 Cyclic Voltammetry

For an ideal capacitor, the ratio $\Delta Q/\Delta V$ (= C) should always remain constant throughout the working potential window. Further information related to this can be acquired by inspecting the i–V response of an electrochemical system. This simple yet powerful observation of a typical current–potential response from a material or device could reveal deeper insight regarding the nature of the occurrence of the charge storage process.

In the case of EDLCs, the cyclic voltammetry curve is a symbolic representation of an almost constant $\Delta Q/\Delta V$ ratio, i.e., the amount of charge held by the material/device remains unchanged throughout the anodic and cathodic scan within the working potential window. However, this is not the case with pseudocapacitive materials, though they produce almost similar cyclic voltammetry curves. In this case, the capacitance does not remain constant throughout the working potential window, which means that the charge held by the material/device fluctuates with variation in the applied potential steps. Since, the ratio, $\Delta Q/\Delta V$, is no longer a constant value, thus the pseudocapacitive nature can be well distinguished from the EDLC. This is also true for hybrid supercapacitors in which both non-faradic (capacitive) and non-capacitive (faradic) processes occur.

One of the conventional methods usually practiced while evaluating the charge storage capacity (in terms of capacitance, C) of a material/device is to integrate the cyclic voltammetry curve, calculate the absolute area, and use the following mathematical expression to obtain the specific capacitance:

Specific capacitance

$$= \frac{\text{Area under the voltammetry curve}}{2 \times \text{active electrode mass} \times \text{potential sweep rate} \times \text{working potential window}}. \quad (3.5)$$

Here, the numerator contains the integrated area for a given cyclic voltammetry curve. In the denominator, we have four different factors, namely, the active electrode mass, potential sweep rate, working potential window, and the number "2."

3.6.2.1 Area Under the Cyclic Voltammetry Curve

Simply integrating the area under a standard cyclic voltammetry curve will not give an accurate measure of the capacitance value. If the CV curve contains redox peaks, then we should get more concerned with the calculation of capacity rather than capacitance. It is to be noted that only a rectangular or quasi-rectangular portion of a CV curve can be taken for the calculation of the capacitive contribution. Additional contributions from redox activities can be quantified from the total capacity.

In practice, the CV curves are not perfectly rectangular; hence, we have to integrate the whole curve using appropriate tools. If provided with the option to calculate either the mathematical area or the absolute area, it is always recommended to go for the absolute area, because opting for mathematical area would take the quadrant signs into consideration and there would be cancellation of the integrated terms due to opposite signs. Care should be taken regarding the baseline and its position/orientation in the cyclic voltammetry plot, while calculating the mathematical/absolute area.

3.6.2.2 Active Electrode Mass

There are various methods to calculate specific capacitance, e.g., we can have volumetric, areal, linear, and gravimetric capacitance values depending on the type of the working electrode used. Areal and gravimetric capacitances are the two mostly used notations to express specific capacitance values. However, there are certain cases where the calculation of areal capacitance can be difficult, e.g., in the case of metal foams (usually taken as the substrates) and several other porous substrates. Gravimetric capacitance, on the contrary, requires the mass of the active material during the calculation of the specific capacitance of a supercapacitor. Having said that, gravimetric calculations are prone to errors too, especially, where lower electrode masses (of the order of milligrams) are involved (this is frequently observed in the case of three-electrode measurements). Since active electrode mass is an important parameter used to determine the specific capacitance, even one order of error term could bring about a tenfold increase in the specific capacitance value. While both the areal and gravimetric capacitance are not absolutely free from errors, they have their own set of advantages when it comes to the design and structure of the working electrode. Nevertheless, gravimetric calculation is often preferred over the areal calculation (along with volumetric and linear calculations), because of the low error terms involved and its compatibility with most of the electrochemical set-ups.

Whether the measurement system is symmetric, asymmetric, or hybrid type, the active mass is calculated by adding the mass present on each electrode side. It should be noted that in the case of asymmetric and hybrid configurations, we usually go for the charge–mass balance calculation; therefore, the mass on each side of the electrode is automatically normalized.

3.6.2.3 Working Potential Window

Standard three-electrode electrochemical measurement is essential in determining the safe working potential window of a metal-ion capacitor. Since it consists of a capacitive electrode and another battery type, therefore, each of the electrodes should be characterized individually in an electrochemical half-cell against a standard reference electrode in order to know the potential range of the electrode material without triggering any side reactions such as hydrogen evolution reaction or oxygen evolution reaction.

Once the potential window is optimized for capacitive and battery components of the hybrid metal-ion device, the potential values obtained should be converted with respect to standard hydrogen electrode potential (as the reference point). The final potential of the hybrid device is the addition of the potentials obtained for each constituent electrode.

3.6.3 Constant Current Charge–Discharge

Although the cyclic voltammetry technique can provide us vital information regarding the charge accumulation process taking place inside a hybrid ion system, calculating several key parameters such as specific capacitance, specific energy, and power, using the CV curve could lead to serious ambiguities. As metal-ion capacitors are energy storage devices, their performance evaluation, therefore, should be done from their charge–discharge profile. This can be done through galvanostatic charge–discharge technique. In fact, this galvanostatic technique can be used to optimize the working potential window of an electrochemical system.

3.7 Principle of Energy Storage in Metal-Ion Capacitors

In a hybrid storage system like metal-ion capacitor, the charge–discharge characteristic is controlled by both non-faradic (capacitive) surface adsorption of charges and non-capacitive (faradic) redox behavior of the battery-type electrode. In case of a capacitive element, the voltage changes almost linearly with the increase in the charge accumulation process. In the case of the battery component, the redox reactions take place at almost a constant potential value (voltage plateau).

In principle, symmetric-type supercapacitors are extremely fast because their electrodes do not go through any redox active process, neither they act as an ion source. Conversely, battery electrodes act as an ion source and are configured in such a way that they can accumulate the desired metal ion through intercalation process. Since supercapacitor electrodes go through charge adsorption/desorption process, the electrolyte becomes the ion source. Therefore, the electrolyte salt gets consumed with each charge–discharge cycle. Conversely, in battery systems, the electrolyte only provides an ion transportation path, where the metal ion migrates in between the two electrodes (i.e., anode and cathode) through the electrolytic medium. Here, the electrolytic content remains intact, unlike supercapacitors. Hence, electrolyte should be considered as an active component of a supercapacitor device. The excessive consumption of the electrolyte results in poor energy density values in the case of supercapacitors.

Combining a capacitive electrode and a battery electrode (impregnated with the metal ions) along with a suitable electrolytic media will readily compensate for the losses that would occur during cycling events. This would significantly boost the energy density of the hybrid device. In fact, the hybrid metal-ion capacitor can be assumed as a special type of asymmetric supercapacitor, where the total energy stored can be expressed as (Li et al. 2018):

$$E = \int V \, dq = \left(V_M - \frac{1}{2} V_C \right) m_B c_B, \tag{3.6}$$

where q is the capacity, and V_M and V_C are the cell's maximum potential and operating potential, respectively. m_B and C_B, respectively, are the mass and specific capacity (in A h g^{-1}) of the battery-type electrode. Similar to the case of an asymmetric supercapacitor, here also mass–charge balancing is important in order to maximize the device performance, as slightest of imbalance in the mass–charge relationship could lead to capacity degradation.

3.8 Performance Metrics for Hybrid Configuration

EDLCs have, for obvious reasons, maximum cycle life among all the electrical energy storage, because they rely purely on surface adsorption mechanism. The cycle life of a metal-ion capacitor lies somewhere in between EDLCs and metal-ion batteries. It is to be noted that the decrease in the cycle life in the case of metal-ion capacitor is well compensated by the impressive energy density values they obtain due to their unique electrode architecture.

Coulombic efficiency depends upon the electrode configuration and the type of electrolyte used in the metal-ion capacitor. A 100% coulombic efficiency, in this case, might not be achievable considering the fact that there are several kinetic limitations. The key role is being played by the electrolyte, which is why clarity regarding the charge balance between the electrodes and the electrolyte as well is essential. The charge accumulated at the surface of the capacitive electrode is proportional to the voltage swing range and can be expressed as:

$$Q_C = m_C C_C V_C, \tag{3.7}$$

where Q_C is the accumulated charge at the surface of the capacitive electrode, m_C is the mass of the capacitive electrode material, C_C is the specific capacitance (F g^{-1}) of the capacitive electrode, and V_C is the maximum working potential window. The maximum charge stored at the surface of carbon is determined by the maximum operational voltage (V_M). Now, the maximum charge stored at the negative electrode is

$$Q_B = m_B C_B, \tag{3.8}$$

where m_B and C_B are the mass and specific capacity (Ah g^{-1}), of the negative electrode material, respectively. For an asymmetric cell, charges at both electrodes should be balanced and also equal the charge of ions consumed from the electrolyte. The maximum charge from ions available in the electrolyte can be expressed as:

$$Q_i = \frac{m_i}{\rho} C_O F, \tag{3.9}$$

where m_i and C_O are the mass and ion concentration of the electrolyte, ρ is the mass density of the electrolyte, and F is the Faraday's constant (96,485 C mol^{-1}).

- *Energy Density*

The capacitance of a symmetric supercapacitor device is half of the capacitance value of a single electrode, because of the series combination of the two electrodes. Therefore, the maximum energy density for a symmetric supercapacitor can be given as (Wang and Zheng 2004):

$$E = \frac{1}{8}C_P V^2. \tag{3.10}$$

If we consider the electrolytic effect, then the expression in Eq. 3.12 changes to

$$E = \frac{1}{8}C_P V^2 \frac{1}{1 + \dfrac{C_P V}{4C_O F}}, \tag{3.11}$$

where the terms $C_P V$ and $C_O F$ are capacitive and capacity contributions, respectively. The capacity contribution is mainly due to the concentration-dependent capacity of the electrolyte.

When the capacitor is in its fully charged state, additional charges stored in positive (q_O) and negative (q_R) electrodes can be described by:

$$q_O = \frac{m_O}{S} \int_{V_0}^{V_{max}} i_O(V)\, dV, \tag{3.12}$$

$$q_R = \frac{m_R}{S} \int_{V_{min}}^{V_0} i_R(V)\, dV. \tag{3.13}$$

where, "s" is the potential sweep rate, "m" and "m" are the mass of the positive and negative electrode, respectively, and "i" and "i" are the current terms associated with the positive and negative electrode, respectively. A more detailed analytical treatment, regarding energy density, electrolyte mass, and constant current approximation, has been provided elsewhere (Wang and Zheng 2004; Zheng 2005).

The final expression for energy density can be provided as follows:

$$
\begin{aligned}
E &= \frac{U_O + U_R}{m_O + m_R} \\
&= \frac{1}{S} \frac{\displaystyle\int_{V_{min}}^{V_0} i_R(V)\,dV \int_0^{V_{max}-V_0} i_O(V)\,dV + \int_{V_0}^{V_{max}} i_O(V)\,dV \int_0^{V_0-V_{min}} i_R(V)\,dV}{\displaystyle\int_{V_0}^{V_{max}} i_R(V)\,dV - \int_{V_{min}}^{V_0} i_O(V)\,dV} \cdot
\end{aligned} \tag{3.14}
$$

where, "E_d" is the energy density, "U_O" and "U_R" are energy stored in the positive and negative electrode, respectively. Since there are so many factors are to be considered, especially in the case of asymmetric and hybrid configurations, detailed measurement and careful analysis of the results (in both half-cell and full-cell configurations) is indispensable to validate and quantify the material characteristics and device performance.

References

Alfaruqi MH, Mathew V, Gim J et al (2015) Electrochemically induced structural transformation in a γ-MnO$_2$ cathode of a high capacity zinc-ion battery system. Chem Mater 27:3609–3620. https://doi.org/10.1021/cm504717p

Amine K, Yasuda H, Yamachi M (1999) β-FeOOH, a new positive electrode material for lithium secondary batteries. J Power Sources 81–82:221–223. https://doi.org/10.1016/S0378-7753(99)00138-X

Aravindan V, Gnanaraj J, Lee Y-S, Madhavi S (2014) Insertion-type electrodes for nonaqueous Li-ion capacitors. Chem Rev 114:11619–11635. https://doi.org/10.1021/cr5000915

Aravindan V, Reddy MV, Madhavi S et al (2011) Hybrid supercapacitor with nano-TiP$_2$O$_7$ as intercalation electrode. J Power Sources 196:8850–8854. https://doi.org/10.1016/j.jpowsour.2011.05.074

Babu B, Shaijumon MM (2017) High performance sodium-ion hybrid capacitor based on Na$_2$Ti$_2$O$_4$(OH)$_2$ nanostructures. J Power Sources 353:85–94. https://doi.org/10.1016/j.jpowsour.2017.03.143

Bauer A, Song J, Vail S et al (2018) The scale-up and commercialization of nonaqueous na-ion battery technologies. Adv Energy Mater 8:1702869. https://doi.org/10.1002/aenm.201702869

Bhat SSM, Babu B, Feygenson M et al (2018) Nanostructured Na2Ti9O19 for hybrid sodium-ion capacitors with excellent rate capability. ACS Appl Mater Interfaces 10:437–447. https://doi.org/10.1021/acsami.7b13300

Bi Z, Paranthaman MP, Menchhofer PA et al (2013) Self-organized amorphous TiO$_2$ nanotube arrays on porous Ti foam for rechargeable lithium and sodium ion batteries. J Power Sources 222:461–466. https://doi.org/10.1016/j.jpowsour.2012.09.019

Chen Z, Augustyn V, Wen J et al (2011) High-performance supercapacitors based on intertwined CNT/V$_2$O$_5$ nanowire nanocomposites. Adv Mater 23:791–795. https://doi.org/10.1002/adma.201003658

Choi HS, Im JH, Kim T et al (2012) Advanced energy storage device: a hybrid BatCap system consisting of battery–supercapacitor hybrid electrodes based on Li$_4$Ti$_5$O$_{12}$–activated-carbon hybrid nanotubes. J Mater Chem 22:16986–16993. https://doi.org/10.1039/C2JM32841K

Choi HS, Kim T, Im JH, Park CR (2011) Preparation and electrochemical performance of hyper-networked Li$_4$Ti$_5$O$_{12}$/carbon hybrid nanofiber sheets for a battery–supercapacitor hybrid system. Nanotechnology 22:405402. https://doi.org/10.1088/0957-4484/22/40/405402

Come J, Naguib M, Rozier P et al (2012) A non-aqueous asymmetric cell with a Ti$_2$C-based two-dimensional negative electrode. J Electrochem Soc 159:A1368–A1373. https://doi.org/10.1149/2.003208jes

Couly C, Alhabeb M, Van Aken KL et al (2018) Asymmetric flexible MXene-reduced graphene oxide micro-supercapacitor. Adv Electron Mater 4:1700339. https://doi.org/10.1002/aelm.201700339

Cui J, Yao S, Ihsan-Ul-Haq M et al (2019) Correlation between Li plating behavior and surface characteristics of carbon matrix toward stable Li metal anodes. Adv Energy Mater 9:1802777. https://doi.org/10.1002/aenm.201802777

Deng Q, Pei J, Fan C et al (2017) Potassium salts of para-aromatic dicarboxylates as the highly efficient organic anodes for low-cost K-ion batteries. Nano Energy 33:350–355. https://doi.org/10.1016/j.nanoen.2017.01.016

Ding J, Hu W, Paek E, Mitlin D (2018) Review of hybrid ion capacitors: from aqueous to Lithium to sodium. Chem Rev 118:6457–6498. https://doi.org/10.1021/acs.chemrev.8b00116

Ding J, Li Z, Cui K et al (2016) Heteroatom enhanced sodium ion capacity and rate capability in a hydrogel derived carbon give record performance in a hybrid ion capacitor. Nano Energy 23:129–137. https://doi.org/10.1016/j.nanoen.2016.03.014

Ding J, Wang H, Li Z et al (2015) Peanut shell hybrid sodium ion capacitor with extreme energy–power rivals lithium ion capacitors. Energy Environ Sci 8:941–955. https://doi.org/10.1039/C4EE02986K

Dong L, Yang W, Yang W et al (2019) Multivalent metal ion hybrid capacitors: a review with a focus on zinc-ion hybrid capacitors. J Mater Chem A 7:13810–13832. https://doi.org/10.1039/C9TA02678A

Dong S, Shen L, Li H et al (2016) Flexible sodium-ion Pseudocapacitors based on 3D Na2Ti3O7 nanosheet arrays/carbon textiles anodes. Adv Funct Mater 26:3703–3710. https://doi.org/10.1002/adfm.201600264

Dong Y, Wu Z-S, Zheng S et al (2017) Ti_3C_2 MXene-derived sodium/potassium titanate nanoribbons for high-performance sodium/potassium ion batteries with enhanced capacities. ACS Nano 11:4792–4800. https://doi.org/10.1021/acsnano.7b01165

Dylla AG, Henkelman G, Stevenson KJ (2013) Lithium insertion in nanostructured TiO2(B) architectures. Acc Chem Res 46:1104–1112. https://doi.org/10.1021/ar300176y

Eftekhari A (2004) Potassium secondary cell based on Prussian blue cathode. J Power Sources 126:221–228. https://doi.org/10.1016/j.jpowsour.2003.08.007

Er D, Li J, Naguib M et al (2014) Ti_3C_2 MXene as a high capacity electrode material for metal (Li, Na, K, Ca) ion batteries. ACS Appl Mater Interfaces 6:11173–11179. https://doi.org/10.1021/am501144q

Gabaudan V, Monconduit L, Stievano L, Berthelot R (2019) Snapshot on negative electrode materials for potassium-ion batteries. Front Energy Res 7:46

Ge P, Hou H, Banks CE et al (2018) Binding MoSe2 with carbon constrained in carbonous nanosphere towards high-capacity and ultrafast Li/Na-ion storage. Energy Storage Mater 12:310–323. https://doi.org/10.1016/j.ensm.2018.02.012

Gummow RJ, Vamvounis G, Kannan MB, He Y (2018) Calcium-ion batteries: current state-of-the-art and future perspectives. Adv Mater 30:1801702. https://doi.org/10.1002/adma.201801702

Han J, Niu Y, Bao S et al (2016) Nanocubic $KTi_2(PO_4)_3$ electrodes for potassium-ion batteries. Chem Commun 52:11661–11664. https://doi.org/10.1039/C6CC06177J

Han J-T, Goodenough JB (2011) 3-V full cell performance of anode framework $TiNb_2O_7$/spinel $LiNi_{0.5}Mn_{1.5}O_4$. Chem Mater 23:3404–3407. https://doi.org/10.1021/cm201515g

Han J-T, Huang Y-H, Goodenough JB (2011) New anode framework for rechargeable lithium batteries. Chem Mater 23:2027–2029. https://doi.org/10.1021/cm200441h

He P, Quan Y, Xu X et al (2017) High-performance aqueous zinc–ion battery based on layered $H_2V_3O_8$ nanowire cathode. Small 13:1702551. https://doi.org/10.1002/smll.201702551

Hosaka T, Shimamura T, Kubota K, Komaba S (2019) Polyanionic compounds for potassium-ion batteries. Chem Rec 19:735–745. https://doi.org/10.1002/tcr.201800143

Jeżowski P, Crosnier O, Deunf E et al (2018) Safe and recyclable lithium-ion capacitors using sacrificial organic lithium salt. Nat Mater 17:167–173. https://doi.org/10.1038/nmat5029

Jung H-G, Venugopal N, Scrosati B, Sun Y-K (2013) A high energy and power density hybrid supercapacitor based on an advanced carbon-coated Li4Ti5O12 electrode. J Power Sources 221:266–271. https://doi.org/10.1016/j.jpowsour.2012.08.039

Kim H, Kim JC, Bianchini M et al (2018) Recent progress and perspective in electrode materials for K-ion batteries. Adv Energy Mater 8:1702384. https://doi.org/10.1002/aenm.201702384

Kim H, Park K-Y, Cho M-Y et al (2014) High-performance hybrid supercapacitor based on graphene-wrapped $Li_4Ti_5O_{12}$ and activated carbon. ChemElectroChem 1:125–130. https://doi.org/10.1002/celc.201300186

Kong L, Zhang C, Wang J et al (2015) Free-standing T-Nb2O5/graphene composite papers with ultrahigh gravimetric/volumetric capacitance for Li-ion intercalation Pseudocapacitor. ACS Nano 9:11200–11208. https://doi.org/10.1021/acsnano.5b04737

Lai C-H, Ashby D, Moz M et al (2017) Designing Pseudocapacitance for Nb_2O_5/carbide-derived carbon electrodes and hybrid devices. Langmuir 33:9407–9415. https://doi.org/10.1021/acs.langmuir.7b01110

Le Z, Liu F, Nie P et al (2017) Pseudocapacitive sodium storage in mesoporous single-crystal-like TiO_2–graphene nanocomposite enables high-performance sodium-ion capacitors. ACS Nano 11:2952–2960. https://doi.org/10.1021/acsnano.6b08332

Leisegang T, Meutzner F, Zschornak M et al (2019) The aluminum-ion battery: a sustainable and seminal concept? Front Chem 7:268

Li B, Zheng J, Zhang H et al (2018) Electrode materials, electrolytes, and challenges in nonaqueous lithium-ion capacitors. Adv Mater 30:1705670

Li C, Deng Q, Tan H et al (2017) Para-conjugated dicarboxylates with extended aromatic skeletons as the highly advanced organic anodes for K-ion battery. ACS Appl Mater Interfaces 9:27414–27420. https://doi.org/10.1021/acsami.7b08974

Li H, Shen L, Pang G et al (2015) $TiNb_2O_7$ nanoparticles assembled into hierarchical microspheres as high-rate capability and long-cycle-life anode materials for lithium ion batteries. Nanoscale 7:619–624. https://doi.org/10.1039/C4NR04847D

Li N, Liu G, Zhen C et al (2011) Battery performance and photocatalytic activity of mesoporous anatase TiO_2 nanospheres/graphene composites by template-free self-assembly. Adv Funct Mater 21:1717–1722. https://doi.org/10.1002/adfm.201002295

Lian P, Dong Y, Wu Z-S et al (2017) Alkalized Ti_3C_2 MXene nanoribbons with expanded interlayer spacing for high-capacity sodium and potassium ion batteries. Nano Energy 40:1–8. https://doi.org/10.1016/j.nanoen.2017.08.002

Liao J, Ni W, Wang C, Ma J (2020) Layer-structured niobium oxides and their analogues for advanced hybrid capacitors. Chem Eng J 391:123489. https://doi.org/10.1016/j.cej.2019.123489

Lim E, Jo C, Kim H et al (2015) Facile synthesis of Nb_2O_5@carbon core–shell nanocrystals with controlled crystalline structure for high-power anodes in hybrid supercapacitors. ACS Nano 9:7497–7505. https://doi.org/10.1021/acsnano.5b02601

Lim E, Kim H, Jo C et al (2014) Advanced hybrid supercapacitor based on a mesoporous niobium pentoxide/carbon as high-performance anode. ACS Nano 8:8968–8978. https://doi.org/10.1021/nn501972w

Liu S, Wang Z, Yu C et al (2013) A flexible TiO_2(B)-based battery electrode with superior power rate and ultralong cycle life. Adv Mater 25:3462–3467. https://doi.org/10.1002/adma.201300953

Liu T, Liu M-M, Zheng X-W et al (2014) Substituent effects on the redox potentials of dihydroxybenzenes: theoretical and experimental study. Tetrahedron 70:9033–9040. https://doi.org/10.1016/j.tet.2014.10.020

Liu W, Hao J, Xu C et al (2017) Investigation of zinc ion storage of transition metal oxides, sulfides, and borides in zinc ion battery systems. Chem Commun 53:6872–6874. https://doi.org/10.1039/C7CC01064H

Liu Z, Huang Y, Huang Y et al (2020) Voltage issue of aqueous rechargeable metal-ion batteries. Chem Soc Rev 49:180–232. https://doi.org/10.1039/C9CS00131J

Lota G, Lota K, Frackowiak E (2007) Nanotubes based composites rich in nitrogen for supercapacitor application. Electrochem Commun 9:1828–1832. https://doi.org/10.1016/j.elecom.2007.04.015

Lukatskaya MR, Mashtalir O, Ren CE et al (2013) Cation intercalation and high volumetric capacitance of two-dimensional titanium carbide. Science (80-) 341:1502 LP–1501505. https://doi.org/10.1126/science.1241488

Ma S, Jiang M, Tao P et al (2018) Temperature effect and thermal impact in lithium-ion batteries: a review. Prog Nat Sci Mater Int 28:653–666. https://doi.org/10.1016/j.pnsc.2018.11.002

Muldoon J, Bucur CB, Gregory T (2014) Quest for nonaqueous multivalent secondary batteries: magnesium and beyond. Chem Rev 114:11683–11720. https://doi.org/10.1021/cr500049y

Naoi K, Ishimoto S, Miyamoto J, Naoi W (2012) Second generation 'nanohybrid supercapacitor': evolution of capacitive energy storage devices. Energy Environ Sci 5:9363–9373. https://doi.org/10.1039/C2EE21675B

Ni J, Yang L, Wang H, Gao L (2012) A high-performance hybrid supercapacitor with $Li_4Ti_5O_{12}$-C nano-composite prepared by in situ and ex situ carbon modification. J Solid State Electrochem 16:2791–2796. https://doi.org/10.1007/s10008-012-1704-9

Peng C-J, Tsai D-S, Chang C, Wei H-Y (2015) The lithium ion capacitor with a negative electrode of lithium titanium zirconium phosphate. J Power Sources 274:15–21. https://doi.org/10.1016/j.jpowsour.2014.10.047

Pérez-Flores JC, Baehtz C, Hoelzel M et al (2012) $H_2Ti_6O_{13}$, a new protonated titanate prepared by Li+/H+ ion exchange: synthesis, crystal structure and electrochemical Li insertion properties. RSC Adv 2:3530–3540. https://doi.org/10.1039/C2RA01134D

Que L-F, Yu F-D, He K-W et al (2017) Robust and conductive $Na_2Ti_2O_5$–x nanowire arrays for high-performance flexible sodium-ion capacitor. Chem Mater 29:9133–9141. https://doi.org/10.1021/acs.chemmater.7b02864

Ramireddy T, Kali R, Jangid MK et al (2017) Insights into electrochemical behavior, phase evolution and stability of Sn upon K-alloying/de-alloying via in situ studies. J Electrochem Soc 164:A2360–A2367. https://doi.org/10.1149/2.0481712jes

Sangster JM (2010) K-P (potassium-phosphorus) system. J Phase Equilibria Diffus 31:68–72. https://doi.org/10.1007/s11669-009-9614-y

Satish R, Aravindan V, Ling WC, Madhavi S (2015) Carbon-coated $Li_3V_2(PO_4)_3$ as insertion type electrode for lithium-ion hybrid electrochemical capacitors: An evaluation of anode and cathodic performance. J Power Sources 281:310–317. https://doi.org/10.1016/j.jpowsour.2015.01.190

Subramanian Y, Veerasubramani GK, Park M-S, Kim D-W (2019) Investigation of layer structured NbSe2 as an intercalation anode material for sodium-ion hybrid capacitors. J Electrochem Soc 166:A598–A604. https://doi.org/10.1149/2.0641904jes

Tarascon J-M, Simon P (2015) New chemistries. In: Electrochemical energy storage. ISTE, London, pp 23–40

Thangavel R, Kaliyappan K, Kang K et al (2016) Going beyond lithium hybrid capacitors: proposing a new high-performing sodium hybrid capacitor system for next-generation hybrid vehicles made with bio-inspired activated carbon. Adv Energy Mater 6:1502199. https://doi.org/10.1002/aenm.201502199

Thangavel R, Kaliyappan K, Kim D-U et al (2017) All-organic sodium hybrid capacitor: a new, high-energy, high-power energy storage system bridging batteries and capacitors. Chem Mater 29:7122–7130. https://doi.org/10.1021/acs.chemmater.7b00841

Thangavel R, Ponraj R, Kannan AG et al (2018) High performance organic sodium-ion hybrid capacitors based on nano-structured disodium rhodizonate rivaling inorganic hybrid capacitors. Green Chem 20:4920–4931. https://doi.org/10.1039/C8GC01987H

Tsai W-Y, Lin R, Murali S et al (2013) Outstanding performance of activated graphene based supercapacitors in ionic liquid electrolyte from −50 to 80°C. Nano Energy 2:403–411. https://doi.org/10.1016/j.nanoen.2012.11.006

Vijayakumar A, Rajagopalan R, Sushamakumariamma AS et al (2015) Synergetic influence of ex-situ camphoric carbon nano-grafting on lithium titanates for lithium ion capacitors. J Energ Chem 24:337–345. https://doi.org/10.1016/S2095-4956(15)60320-5

Wang X, Zheng JP (2004) The optimal energy density of electrochemical capacitors using two different electrodes. J Electrochem Soc 151:A1683. https://doi.org/10.1149/1.1787841

Wang F, Wang X, Chang Z et al (2015) A quasi-solid-state sodium-ion capacitor with high energy density. Adv Mater 27:6962–6968. https://doi.org/10.1002/adma.201503097

Wang R, Wang S, Peng X et al (2017) Elucidating the intercalation Pseudocapacitance mechanism of MoS_2–carbon monolayer interoverlapped superstructure: toward high-performance sodium-ion-based hybrid Supercapacitor. ACS Appl Mater Interfaces 9:32745–32755. https://doi.org/10.1021/acsami.7b09813

Wang M, Jiang C, Zhang S et al (2018) Reversible calcium alloying enables a practical room-temperature rechargeable calcium-ion battery with a high discharge voltage. Nat Chem 10:667–672. https://doi.org/10.1038/s41557-018-0045-4

Wang P, Ye H, Yin Y-X et al (2019a) Fungi-enabled synthesis of ultrahigh-surface-area porous carbon. Adv Mater 31:1805134. https://doi.org/10.1002/adma.201805134

Wang Y, Zhang Z, Wang G et al (2019b) Ultrafine Co2P nanorods wrapped by graphene enable a long cycle life performance for a hybrid potassium-ion capacitor. Nanoscale Horizons 4:1394–1401. https://doi.org/10.1039/C9NH00211A

Wu N, Yao W, Song X et al (2019) A calcium-ion hybrid energy storage device with high capacity and long cycling life under room temperature. Adv Energy Mater 9:1803865. https://doi.org/10.1002/aenm.201803865

Xie Y, Dall'Agnese Y, Naguib M et al (2014) Prediction and characterization of MXene nanosheet anodes for non-Lithium-ion batteries. ACS Nano 8:9606–9615. https://doi.org/10.1021/nn503921j

Xu C, Du H, Li B et al (2009) Reversible insertion properties of zinc ion into manganese dioxide and its application for energy storage. Electrochem Solid-State Lett 12:A61. https://doi.org/10.1149/1.3065967

Xu N, Sun X, Zhang X et al (2015) A two-step method for preparing $Li_4Ti_5O_{12}$–graphene as an anode material for lithium-ion hybrid capacitors. RSC Adv 5:94361–94368. https://doi.org/10.1039/C5RA20168C

Xu W, Wang Y (2019) Recent progress on zinc-ion rechargeable batteries. Nano-Micro Lett 11:90. https://doi.org/10.1007/s40820-019-0322-9

Xu W, Zhao K, Wang Y (2018) Electrochemical activated MoO2/Mo2N heterostructured nanobelts as superior zinc rechargeable battery cathode. Energy Storage Mater 15:374–379. https://doi.org/10.1016/j.ensm.2018.06.028

Yan M, He P, Chen Y et al (2018) Water-lubricated intercalation in $V_2O_5 \cdot nH_2O$ for high-capacity and high-rate aqueous rechargeable zinc batteries. Adv Mater 30:1703725. https://doi.org/10.1002/adma.201703725

Yang H, Li J-S, Zeng X (2018a) Correlation between molecular structure and interfacial properties of edge or basal plane modified graphene oxide. ACS Appl Nano Mater 1:2763–2773. https://doi.org/10.1021/acsanm.8b00405

Yi R, Chen S, Song J et al (2014) High-performance hybrid supercapacitor enabled by a high-rate Si-based anode. Adv Funct Mater 24:7433–7439. https://doi.org/10.1002/adfm.201402398

Yin J, Qi L, Wang H (2012) Sodium titanate nanotubes as negative electrode materials for sodium-ion capacitors. ACS Appl Mater Interfaces 4:2762–2768. https://doi.org/10.1021/am300385r

Zhang C, Lv W, Xie X et al (2013a) Towards low temperature thermal exfoliation of graphite oxide for graphene production. Carbon N Y 62:11–24. https://doi.org/10.1016/j.carbon.2013.05.033

Zhang N, Cheng F, Liu Y et al (2016) Cation-deficient spinel $ZnMn_2O_4$ cathode in $Zn(CF_3SO_3)_2$ electrolyte for rechargeable aqueous Zn-ion battery. J Am Chem Soc 138:12894–12901. https://doi.org/10.1021/jacs.6b05958

Zhang R, Ling C (2016) Status and challenge of Mg battery cathode. MRS Energy Sustain 3:E1. https://doi.org/10.1557/mre.2016.2

Zhang L, Wilkinson DP, Chen Z, Zhang J (2018a) Lithium-ion supercapacitors: fundamentals and energy applications. CRC Press, Boca Raton

Zhang P, Zhao X, Liu Z et al (2018b) Exposed high-energy facets in ultradispersed sub-10 nm SnO_2 nanocrystals anchored on graphene for Pseudocapacitive sodium storage and high-performance quasi-solid-state sodium-ion capacitors. NPG Asia Mater 10:429–440. https://doi.org/10.1038/s41427-018-0049-y

Zhang Z, Li M, Gao Y et al (2018c) Fast potassium storage in hierarchical $Ca_{0.5}Ti_2(PO_4)_3$@C microspheres enabling high-performance potassium-ion capacitors. Adv Funct Mater 28:1802684. https://doi.org/10.1002/adfm.201802684

Zhang W, Liu Y, Guo Z (2019) Approaching high-performance potassium-ion batteries via advanced design strategies and engineering. Sci Adv 5:eaav7412. https://doi.org/10.1126/sciadv.aav7412

Zhang X, Wang L, Liu W et al (2020a) Recent advances in MXenes for lithium-ion capacitors. ACS Omega 5:75–82. https://doi.org/10.1021/acsomega.9b03662

Zhang Y, Jiang J, An Y et al (2020b) Sodium-ion capacitors: materials, mechanism, and challenges. ChemSusChem 13:2522–2539. https://doi.org/10.1002/cssc.201903440

Zhao X, Johnston C, Grant PS (2009) A novel hybrid supercapacitor with a carbon nanotube cathode and an iron oxide/carbon nanotube composite anode. J Mater Chem 19:8755–8760. https://doi.org/10.1039/B909779A

Zhao X, Zhang Q, Chen C-M et al (2012) Aromatic sulfide, sulfoxide, and sulfone mediated mesoporous carbon monolith for use in supercapacitor. Nano Energy 1:624–630. https://doi.org/10.1016/j.nanoen.2012.04.003

Zhao Q, Wang J, Lu Y et al (2016) Oxocarbon salts for fast rechargeable batteries. Angew Chemie Int Ed 55:12528–12532. https://doi.org/10.1002/anie.201607194

Zhao X, Cai W, Yang Y et al (2018a) MoSe$_2$ nanosheets perpendicularly grown on graphene with Mo–C bonding for sodium-ion capacitors. Nano Energy 47:224–234. https://doi.org/10.1016/j.nanoen.2018.03.002

Zhao X, Wang H-E, Yang Y et al (2018b) Reversible and fast Na-ion storage in MoO$_2$/MoSe$_2$ heterostructures for high energy-high power Na-ion capacitors. Energ Storage Mater 12:241–251. https://doi.org/10.1016/j.ensm.2017.12.015

Zhao X, Zhao Y, Liu Z et al (2018c) Synergistic coupling of lamellar MoSe$_2$ and SnO$_2$ nanoparticles via chemical bonding at interface for stable and high-power sodium-ion capacitors. Chem Eng J 354:1164–1173. https://doi.org/10.1016/j.cej.2018.08.122

Zhao J, Wang G, Hu R et al (2019) Correction: ultrasmall-sized SnS nanosheets vertically aligned on carbon microtubes for sodium-ion capacitors with high energy density. J Mater Chem A 7:13790. https://doi.org/10.1039/C9TA90127B

Zheng JP (2005) Theoretical energy density for electrochemical capacitors with intercalation electrodes. J Electrochem Soc 152:A1864. https://doi.org/10.1149/1.1997152

Zhu Y-E, Yang L, Zhou X et al (2017) Boosting the rate capability of hard carbon with an ether-based electrolyte for sodium ion batteries. J Mater Chem A 5:9528–9532. https://doi.org/10.1039/C7TA02515G

Chapter 4
Advantages and Challenges

Abstract The prospects associated with the development of hybrid metal-ion capacitors are promising, though the cost economy and commercial viability are going to pose few stiff challenges in this case. However, these next generation storage technologies are critical to sustain the rapidly depleting fossil fuel reserves and to establish the renewable energy sources as the potential alternatives. Since there are many technological aspects associated with these hybrid metal-ion capacitors, here we briefly discuss about the technological and economic challenges that have to be overcome before these disruptive storage techniques can be realized in practice.

Hybrid energy storage system integrates two or more energy storage technologies with complementary operating characteristics (here, batteries and supercapacitors). These systems typically have one component (capacitive) dedicated to cover high power loads, transients and fast transients, and fast load fluctuations. It is therefore characterized by a fast response time, high efficiency, and high cycle lifetime. And, the other component (battery) will deal with high energy storage with a low self-discharge rate and lower energy-specific installation costs. In addition to the benefits listed above, hybrid systems should also have the ability to operate under dynamic conditions and to have a high system efficiency, resulting in fewer thermodynamic losses. The integrative approach will result in an enhanced storage system with much longer lifespan.

Hybrid systems enhance the overall effectiveness of the existing power systems through reduced cycling and changes in output, resulting in lower emissions and lighter footprint. For applications that require storage to cover short-term power fluctuations such as renewables ramping/smoothing, a technology that is suited for high power operations leads to improved system efficiency. Conversely, using the same hybrid system for low-power, long-duration applications such as backup

© The Author(s), under exclusive license to Springer Nature Switzerland AG 2020
A. K. Samantara, S. Ratha, *Metal-Ion Hybrid Capacitors for Energy Storage*,
SpringerBriefs in Energy, https://doi.org/10.1007/978-3-030-60812-5_4

power will not compromise its overall health. Many of the energy storage industries have just begun exploring grid-scale hybrid solutions, and devices like batteries, supercapacitors, and metal-ion capacitors are to play a significant role in it, because they act as clean energy sources. Also, they can be handled with ease and do not require frequent transportation, and devices like supercapacitors and metal-ion capacitors do not even need maintenance, say, for at least 10,000–20,000 charge–discharge cycles. Nevertheless, there are few technological and economic challenges to this concept of metal-ion capacitors.

4.1 Technological Challenges

The development of an advanced energy storage technology, e.g., metal-ion capacitors, requires top-notch innovation and breakthrough in capacity, long lifespan, low cost, and high security for electrochemical energy storage. And also, a physical storage technology with high efficiency and low cost is required. Although combining both battery and supercapacitor components would substantially improve the balance between energy and power, it will require a steadfast approach from both academic and industrial sectors, to address the technological challenges involved.

4.2 Economic Challenges

Currently, global energy storage industries are still facing inherent challenges, e.g., lack of policy support, high cost, unclear application value, and other issues. Two aspects should be considered in the future: it is necessary to propose energy storage system solutions in terms of technological capabilities, active participations from the industries, and at the user level including general public, researchers, and economic and social organizations. High-grade research activities and time-bound applications of energy storage solutions are essential in establishing a sustainable development model and achieving commercial operation of energy storage.

Chapter 5
Summary and Conclusion

Abstract Batteries and supercapacitors have their share of advantages and limitations. We must realize that there could be no singular solution to address the energy crisis and to develop/build a sustainable energy storage framework. In this regard, we can only hope that the current energy supply chain will be able to cope with the growing demand for energy. Therefore, we must look for reinforcements, not alternatives, and since most of our energy requirements are met in the form of electrical energy, its storage through the implementation of direct and/or indirect methods is inevitable. Batteries have high energy densities, but they fail when it comes to generating power, enough to run transportation vehicles. Also, their capacity fades quite fast, and recycling is an emerging issue. Supercapacitors can be advantageous, because they can deliver much higher power densities and can run for more than 1,00,000 charge–discharge cycles. However, lackluster energy density values limit the application of supercapacitors in storage devices. If both of these technologies can be combined, and if we just can do few basic energy and power calculations, then we can have both high power (from supercapacitors) and energy (from batteries). Nevertheless, combining these two different techniques is not straightforward and requires strategic integrative approach to achieve maximum efficiency. One such attempt is to combine/integrate both the battery and supercapacitor chemistries to form a hybrid system, where both fast charging and long-duration storage can be afforded.

Batteries and supercapacitors have their share of advantages and limitations. We must realize that there could be no singular solution to address the energy crisis and to develop/build a sustainable energy storage framework. In this regard, we can only hope that the current energy supply chain will be able to cope with the growing demand for energy. Therefore, we must look for reinforcements, not alternatives,

© The Author(s), under exclusive license to Springer Nature Switzerland AG 2020   97
A. K. Samantara, S. Ratha, *Metal-Ion Hybrid Capacitors for Energy Storage*,
SpringerBriefs in Energy, https://doi.org/10.1007/978-3-030-60812-5_5

and since most of our energy requirements are met in the form of electrical energy, its storage through the implementation of direct and/or indirect methods is inevitable. Batteries have high energy densities, but they fail when it comes to generating power, enough to run transportation vehicles. Also, their capacity fades quite fast, and recycling is an emerging issue. Supercapacitors can be advantageous, because they can deliver much higher power densities and can run for more than 1,00,000 charge–discharge cycles. However, lackluster energy density values limit the application of supercapacitors in storage devices. If both of these technologies can be combined, and if we just can do few basic energy and power calculations, then we can have both high power (from supercapacitors) and energy (from batteries). Nevertheless, combining these two different techniques is not straightforward and requires strategic integrative approach to achieve maximum efficiency. One such attempt is to combine/integrate both the battery and supercapacitor chemistries to form a hybrid system, where both fast charging and long duration storage can be afforded.

At domestic level, storage requirements are managed by secondary batteries, with lithium-ion batteries grabbing a major share. After almost 30 years of commercialization, lithium ion is now facing an ineluctable issue, i.e., scarcity of lithium source. However, new and emergent concepts such as sodium-, potassium-, and other multivalent metal-ion-based technologies are peaking up. This does not mean that they shall be able to replace the current lithium-ion cells. But they can readily support it by reducing the stress arising from global energy demand and to promote renewable energy sources, for sustainable growth. The concept of metal-ion capacitor has grown from the idea of asymmetric supercapacitor devices, and it aims to integrate a battery component and a supercapacitor component into a single device (hybrid metal-ion capacitor) to yield faster charging and longer storage performance.

In this Brief, we have tried to provide few basics and fundamentals of supercapacitors and batteries (not elaborated due to vastness of the topic), where they fail and how can we make them go through it. Also, a concise detail of hybrid storage and various types of metal-ion capacitor technologies has been provided to clearly understand the basic working mechanism and to highlight the key components that affect the device performance metrics.

Index

© The Author(s), under exclusive license to Springer Nature Switzerland AG 2020   99
A. K. Samantara, S. Ratha, *Metal-Ion Hybrid Capacitors for Energy Storage*,
SpringerBriefs in Energy, https://doi.org/10.1007/978-3-030-60812-5

Printed in the United States
by Baker & Taylor Publisher Services